·新世纪侨校高等教育精品教材·

聚合物成型

工艺与技术

Polymer Molding Process and Technology

林志丹　编著

暨南大學出版社
JINAN UNIVERSITY PRESS

中国·广州

图书在版编目（CIP）数据

聚合物成型工艺与技术/林志丹编著．—广州：暨南大学出版社，2014.12
ISBN 978－7－5668－1034－2

Ⅰ.①聚…　Ⅱ.①林…　Ⅲ.①高聚物—成型加工　Ⅳ.①TQ316

中国版本图书馆 CIP 数据核字（2014）第 110179 号

出版发行：暨南大学出版社

地　址：中国广州暨南大学
电　话：总编室（8620）85221601
营销部（8620）85225284　85228291　85228292（邮购）
传　真：（8620）85221583（办公室）　85223774（营销部）
邮　编：510630
网　址：http：//www.jnupress.com　http：//press.jnu.edu.cn

排　版：广州市天河星辰文化发展部照排中心
印　刷：佛山市浩文彩色印刷有限公司

开　本：787mm×1092mm　1/16
印　张：8
字　数：200 千
版　次：2014 年 12 月第 1 版
印　次：2014 年 12 月第 1 次

定　价：26.80 元

前　言

近些年来，双语教学在很多国家或地区日益发展起来。当然，不同的国家或地区双语所指的具体内容也有所不同。在我国大陆地区，最早的双语概念主要指各种不同的少数民族语言和汉语普通话，因此，当时的双语研究和教学主要集中在少数民族学习汉语的问题上。随着我国国际地位的提升，在国际上的影响逐渐扩大，近年来，国内又掀起了双语教学的热潮，这时的双语主要是指汉语和英语。双语教学不是简单地为了强化英语教学，不是通过其他学科达到掌握英语的目的。就目前情况看，由于双语教学刚刚起步，受教材、师资和学生现有英语水平的限制，如果一说开展双语教学就要全部用英语进行各个学科的知识教学，这显然是不切合实际的。经过广泛的调研、深入的讨论，我们在双语教材的内容、形式等方面都充分考虑了现有双语教学的不足。本书以英文为主，并适当辅以汉语的解释和说明。这种穿插型模式的教材更有利于学生理解与掌握。

本书作者在认真学习国内外高分子加工著作的基础上，总结并汲取塑料加工科学的新成果，力图使本书在科学思路和内容编排上具有以下特点：①作为基础性教材，内容尽可能系统和全面。通过介绍塑料成型的基本概念、经典理论、主要设备与塑化技术，使读者获得扎实的塑料加工基础知识；②编排合理，体现科学思路。在对基本概念和成型原理进行介绍的基础上，主要按压制成型、挤出成型、注射成型、压延成型、中空吹塑和热成型的分类介绍塑料加工原埋和技术；③紧跟学科发展的前沿。介绍传统塑料加工成型理论的同时结合了近年来的最新研究成果和研究方法。

由于此模式还处于试验、探索阶段，还需要在实践中不断发展和完善，希望读者对双语教材的出版提出建议、意见，以便我们在今后的工作中逐步改进、完善。

编　者

2014年3月

CONTENTS

Chapter 1
Pressing Forming

压制成型

1.1 Basic Concepts

基本概念

Press forming is a kind of technique which mainly relies on external pressure to achieve press forming of material modeling. According to the articles about pressing, press forming can be divided into compression molding and laminate molding. According to whether the state of material change or not in the process of forming, press forming can be divided into ordinary press forming and powder press forming. Powder press forming refers to a kind of press forming which uses powder material to form and makes powder material always maintain solid during the forming process, besides, its forming material is usually only ware (坯件), and it is in need of a follow-up sintering process to obtain the final products of press forming. Powder press forming will not be introduced in this chapter, but in the part of "powder press molding and sinter molding (烧结成型)" we will mention it. Ordinary press forming refers to the material of forming process (the common compression molding) or part of the material (such as laminate molding bonding material) which will cause, or partly cause the process of state changing of solid—liquid—solid state. Compacting process typically involves physical problems and chemical reactions about press forming.

This chapter will be focused on introducing the mold, principle, technique and development of the common compression molding, and will give a brief introduction to laminate molding, on the basis of introducing forming equipments.

1.2 Press Forming Devices

压制成型设备

The core equipment of press forming is press. Compression molding needs to use die (模具) and some kinds of press forming need auxiliary equipment. In this chapter, we will only introduce the press.

1.2.1 Types of Press

液压机

Press is known as a pressure machine which uses a plate to put pressure on plastics during the forming process. According to the transmission styles, press can be divided into hydraulic press and mechanical press. The more widely used one is the hydraulic press and the commonly used hydraulic press is the oil hydraulic press. Hydraulic press can be divided into general hydraulic press and multi-layer hydraulic press.

1. General Hydraulic Press

柱式液压机/框式液压机

According to the characteristics of the frame, general hydraulic press can be divided into column hydraulic press and frame hydraulic press . According to the setting position of working cylinder, the general hydraulic press has pressurizing-down style and pressurizing-up style. General hydraulic press has three kinds of operation methods: manual method, semi-automatic method and automatic method respectively. The specifications of general hydraulic press are 450 kN, 630 kN, 1,000 kN, 1,600 kN, 2,500 kN, 3,000 kN, 5,000 kN, etc.

工作油缸
柱塞

The structure of the pressurizing-up style hydraulic press is shown in Fig. 1 - 1(a). The working cylinder is designed on the top of the hydraulic press, and the lower plate is fixed. The main pressure plunger pushes the upper plate to press from the top to the bottom, depending on the rise and fall of the upper plate to complete the opening and closing of the mold and add pressure to the material.

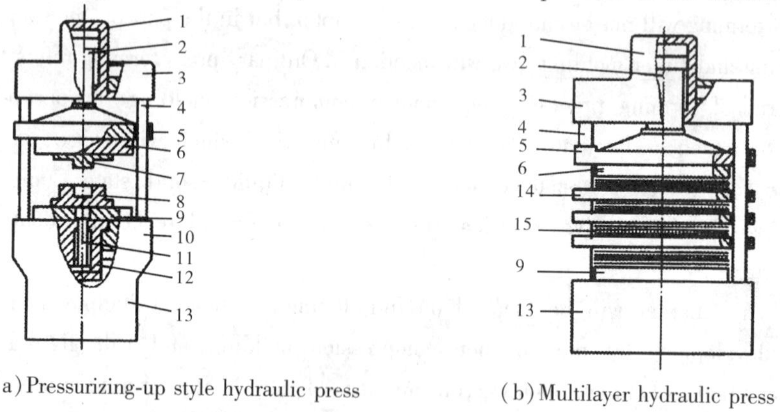

(a) Pressurizing-up style hydraulic press　　(b) Multilayer hydraulic press

Fig. 1 - 1　pressurizing-up style hydraulic press and multilayer hydraulic press structural diagram

(1—master cylinder; 2—master cylinder piston; 3—upper beam; 4—pillar; 5—activities board; 6—the upper template; 7—plug; 8—mold; 9—the lower template; 10—the machine; 11—ejection cylinder piston; 12—ejection cylinder; 13—engine base; 14—layered activities board; 15—laminate plastic)

The position of working cylinder of pressurizing-down style hydraulic press is located at the foot of the hydraulic press and the upper press plate is fixed. The main pressure plunger is located under the lower pressure plate, and the main pressure piston pushes the lower plate to go up , so as to realize the closing and opening of the die.

2. Multilayer Hydraulic Press

Multilayer hydraulic press is a specialized hydraulic press, between the upper compressor and the lower compressor of press. There exists many heating plates which can move up and down. Its structure is shown in Fig. 1 – 1(b). Multilayer hydraulic press can simultaneously form three to dozens of layers of different slabs, thus its production efficiency is high. Multilayer hydraulic press can produce laminate, rigid PVC lumber, foam plastic product, etc. The specifications of multilayer hydraulic press are 10NM, 15NM, 20NM, 25NM, 30NM, etc.

加热板

聚氯乙烯/发泡塑料制品

1.2.2 The Parameters of Press

参数

The main parameters of press are the nominal pressure , the plunger diameter, the size of pressure plate and the working stroke. The size of pressure plate determines the scale of products which the press can make. The working stroke determines the height of the mold or products. The nominal pressure of hydraulic press is the main parameter which can demonstrate the pressing capability of the press. The specification of hydraulic press is usually expressed by nominal pressure. Nominal pressure—F can be calculated as follow:

公称力

$$F = P_L \times \frac{\pi D^2}{4}$$

In the formula, D is the diameter of the oil hydraulic plunger, P_L is the greatest pressure that press can bear.

Chapter 2
Extrusion Molding

挤出成型

2.1 Outline

概述

热塑性塑料

Extrusion molding is also known as extrusion or plastic extrusion. In thermoplastic processing field, extrusion is a more changeable, broadly used processing approach which accounts for a large proportion in the plastics processing. The extruded products are all extendable material with the fixed cross-sections, such as a tube, a plate, a thread, a film, the coating of wire and a cable, etc. But the extrusion is limited in thermoplastic processing. Apart from profile extrusion, it can also be used for mixing, plasticization, granulation, coloring and so on.

螺杆
摩擦

Pipe extrusion molding technique is shown in Fig. 2 - 1. Plastics is placed into the hopper. With the help of the rotation of the screw , the plastics enters into the feeding tube. Due to the heating of the hopper and the shear friction heat bot from the plastics itself and between plastics and the equipment, the plastics is melted and becomes the fluid state. At the same time, the melted plastics is stirred by the screw, dispersed uniformly and advanced continually. Finally, after the melted plastics is shaped by the nose body, stereotyped by molding unit and cooled, the product comes out. As long as you replace the die head and the corresponding auxiliary equipments, different products are obtainable.

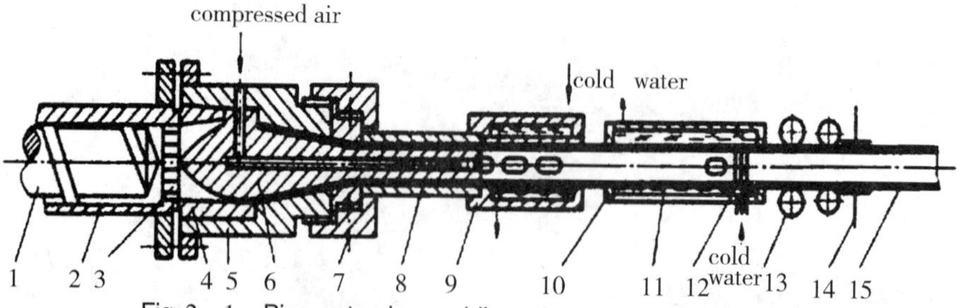

Fig. 2 - 1 Pipe extrusion molding process schematic diagram

(1—screw; 2—barrel; 3—die cone; 4—interface sets; 5—nose body; 6—core-rod; 7—adjusting screws; 8—die; 9—sizing sleeve; 10—cooling tank; 11—chain; 12—plug; 13—traction device; 14—clamping device; 15—plastic tubes)

From the above phenomena, extrusion process seems complicated superficially, but it is divided into two phases theoretically. The first phase is to plasticize the plastics (i. e. become the viscous fluid), and the plastics is forced through a special mouth die shaped to form a continuum whose cross section is similar to the shape of mouth die under pressure. The second phase is to make the extruded continuous body leave the plastic state and become solid by adopting a proper method, and obtain the desired product.

黏性流体
口模

The plasticization of plastics can not only use heating to make plastics melted (the dry method), but also use solvent to soft plastics sufficiently (wet method), thus, according to the difference between the approaches of plasticization of plastic, the extrusion technique can be divided into dry method and wet method. For dry method, its characteristics are that the plasticizing and pressing can be conducted in the same device and the stereotype of its products can be completed only by using cooling. However, it is easy to arouse the phenomena of uneven plasticization or overheat in the process of heating due to the poor heat-transfer ring capability of the plastics. For wet method, the plastics needs to be soaked in the solvent for a long time. Even though the plasticizing is uniform and the overheating can be avoided, the plasticizing and pressing should be divided into two separate processes. Besides, the setting treatment must adopt a troublesome solvent removal process and the solvent recycling should also be taken into consideration. Based on these disadvantages, it only applies to plastic extrusion of cellulose nitrate and minority of cellulose acetate .

硝酸纤维素
醋酸纤维素

During the extrusion process, it can be divided into continuous method and intermittent method according to different plastic pressurizing methods. The equipment of the former way is a screw extruder; the latter is a plunger-type extruder. The types of the screw extruder includes single-screw extruder and multi-screw extruder, and the single-screw extruder is used more frequently. The main components of the plunger-type extruder are a sleeve and a hydraulically controlled plunger. When operating, firstly, we should pour a batch of plasticizing plastics into the sleeve, then extrude the plastics through extrusion die with the pressure of plunger, and remove the plunger for the next operation after squeezing the plastics. The biggest advantage of the plunger-type extruder is its ability to offer plastics a greater pressure, while its obvious disadvantage is the discontinuous operation as well as its requiring pre-plasticizing material beforehand. Thus, it has fewer applications. It is suitable for the extrusion of polytetrafluoroethylene, ultrahigh

挤出机
聚四氟乙烯

molecular weight polyethylene and so on.

In conclusion, the majority of plastic extrusion is thermoplastic plastics which adopts the dry method to squeeze continuously.

2.2 The Basic Structure of the Single-screw Extruder

阿基米德螺杆

The single-screw extruder consists of one Archimedes screw rotating in the heating sleeve. The size of the single-screw extruder is usually expressed by the screw diameter and its basic structure mainly includes transmission device, charging device, the sleeve, screw, etc., as illustrated in Fig. 2 – 2.

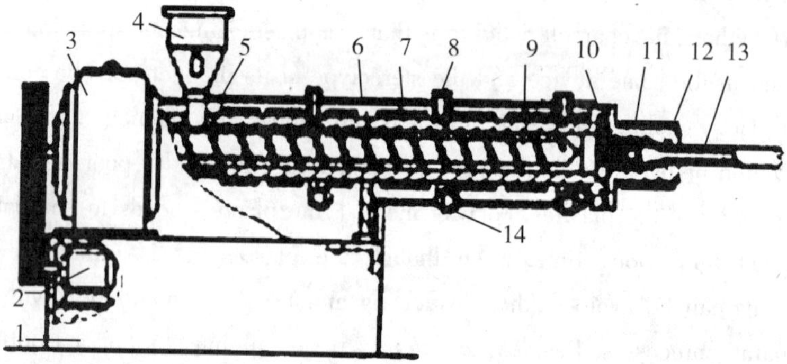

Fig. 2 – 2 The schematic diagram of the single-screw extruder structure

传动装置

2.2.1 Transmission Device

轴承

The transmission device is the part which drives the screw to rotate, and it is usually composed of the electric motor, the deceleration mechanism, and the bearing. In order to ensure the quality of products, there are several basic requirements for the transmission device. Firstly, in normal operating conditions, no matter whether the load of the screw changes or not, the rotational speed of the screw should remain unchanged. Because if the screw speed changes in the extrusion process, it will cause the pressure fluctuation of the plastic material flow. Therefore, it will be difficult to maintain the stability of the product quality. Secondly, the screw speed of the same extruder can be adjusted in order to extrude different extruded products or different plastics. To meet these requirements, it had better use infinitely variable speed transmission device. There are three ways to achieve infinitely variable speed. ① Commutator motor or DC motor, which is both

整流子式电动机

a drive device and a shifting device. ②Mechanical friction transmission drove by constant speed motor, such as the CVT apparatus of a gear transmission. ③Motor-driven oil pump sending the oil to the hydraulic motor and changing the fuel discharge of the pump in order to change the speed of the extruder screw.

Transmission device should have a good lubrication system and a rapid braking device. 润滑系统

2.2.2 Feeding Device 加料装置

The kinds of lastics which are used to be processed are various, such as granular plastic, powdery plastic, ribbon plastic and so on. The feeding device generally applies a hopper whose capacity is able to accommodate for materials at least 1h. A hopper should have devices used to cut off the material flow, calibrate material quantity and drop remnant. A better hopper has devices such as timing, feeding quantitatively, inherent drying or preheating. Besides, it also adopts feeding under reduced pressure, which is known as vacuum feeding device, and is particularly suitable for the hygroscopic plastics processing and powder material processing. With the improvement of extrusion equipment and technique, the supply amount of power feeding is more and more. Powder and pellets are gravity-fed from the hopper into the feeding hole, but with change of the height of the material layer, it may cause the change of the feed rate at the same time. It may produce the phenomenon of " bridging ", which may cause material shortage. A setting agitator or screw forced feeder conveyor in feeding device can overcome this disadvantage (shown in Fig. 2 – 3). The feeding hole can be rectangular or round but the rectangular is used more widely. Its long sides are parallel to the axis, with 1 – 1.5 times the length of the screw diameter and 7 ° – 15 ° the tilt angle on the feeding side. Around feeding hole, it should be provided with the cooling jacket (shown in Fig. 2 – 4) to eliminate the high-temperature cylinder from transferring heat to the hopper and avoid plastics in hopper being tacky because of rising temperature, thus resulting in uneven feeding or blocked stream.

料斗

架桥

矩形

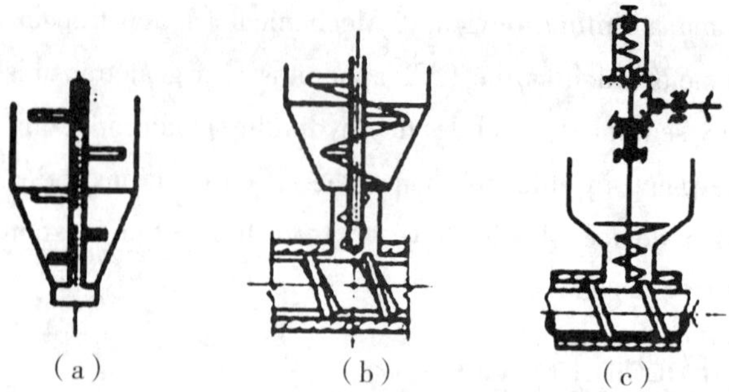

Fig. 2－3　Forced feeder

[(a) mixing feeder; (b) screw feeder; (c) feeding quantity variable feeder with protective effect]

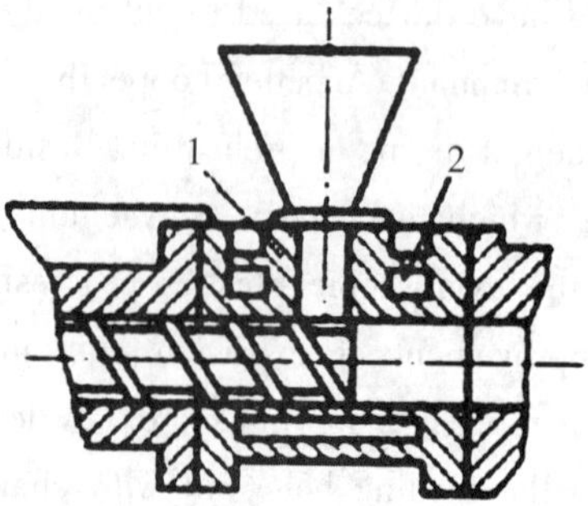

Fig. 2－4　Cooling device of feeding bucket seat

(1—hopper block; 2—cooling water channel)

料筒

2.2.3　Sleeve

Sleeve is one of the main components of the extruder, where the process of plastic plasticizing and pressurizing are carried out. During extrusion, the pressure of cylinder is up to 55 MPa, and the operating temperature is generally 180 ℃ － 300 ℃, so the sleeve can be regarded as pressure and heating container. The material manufactured the sleeve should be of high strength, fastness, wear and corrosionresistance, which is generally composed of steel shell and alloy steel liner. Its outside sets the part of heating and cooling device, each with a thermocouple and automatic meter respectively. The heating methods include resistance heating and induction heating. The latter could heat better and cool more conveniently, but it is costly.

耐腐蚀性/合金

热电偶

Extruder can also be heated by usable oil and steam, and it owns the advantage of uniform heating under a certain range of temperature, but it is rarely used now because of the complicated apparatus, and the limited temperature control range . There will be a risk of increasing the products contamination. 污染

Usually the sleeve is also provided with a cooling system, whose main role is to prevent the plastics from overheating, or rapid cooling when stopping in case of resin degradation or decomposition. The sleeve is generally cooled by air or water. 降解 Some of the fins of extruder sleeve or heater aim to increase the efficiency of the air cooling. As for the cooling efficiency, cooling with cold water through copper tubes embedded in the sleeve is effective. However, it easily leads to rapid cooling, occurring fouling, rusting and other negative phenomena.

2.2.4 Screw

螺杆

The screw is the key part of the extruder. Through its rotation, the plastics within the sleeve will move, generating a supercharger and a portion of the heat (friction heat). Screw geometric parameters, such as diameter, aspect ratio, each segment length ratio, the spiral groove depth and so on, have a significant impact 螺槽深度 on the operating characteristics of the screw. Therefore, the basic parameters of the screw and its role will be introduced briefly in the following passage. The screw structure is shown in Figure 2 - 5.

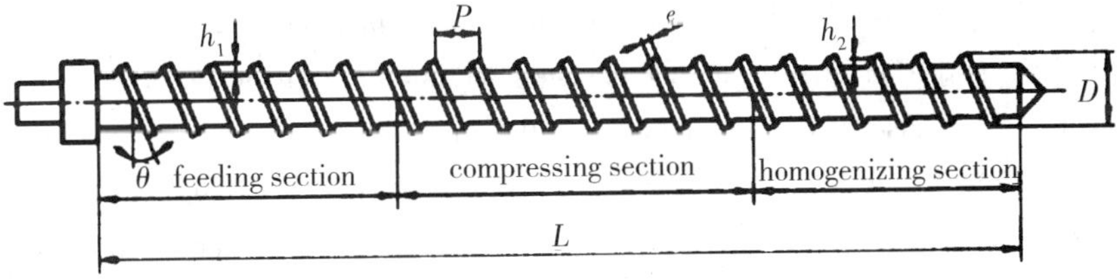

Fig. 2 - 5 Screw

(h_1—feeding section spiral groove depth; h_2—homogenizing section spiral groove depth; D—screw diameter; θ—helix angle; L—screw length; e—screw flight width; P—pitch)

1. Screw Diameter (D) and Aspect Ratio (L / D)

Screw diameter is one of the basic parameters of the screw, commonly used to represent the size of the extrusion. Users should select screw diameter according to the shape and size of the product and other requirements of productivity. In addition, other parameters such as the screw length, the depth of groove and the width of the screw flight are closely related to the diameter, which are mostly 螺棱 represented by diameter ratio.

长径比

剪切

Another important parameter to characterize screw is the ratio of effective length(L)and diameter,i. e. the aspect ratio (L / D). If the screw is only taken as the way of materials transportation, the aspect ratio of the screw mainly determines its volumetric capacity. Besides,the aspect ratio also affects the rate of heat transition from the sleeve wall to the material,thus affects the heat generated by shearing the energy input and the ratio of the quantity of power and the extrusion amount. Therefore,increasing the aspect ratio can prolong the residence time of the plastics on the screw to plasticize more uniformly and can accelerate the screw speed to increase the extrusion amount. Currently,the aspect ratio has been increased,but a too long screw will bring some difficulties for manufacturing and assembling. Generally,aspect ratio is around 25.

2. Functions of Each Segments of Screw

螺纹角

螺距

According to the operation of the plastic on the screw,it can be divided into feeding segment,compression and metering segment. The general screw,sometimes referred as a standard screw or a metering type screw, has a thread angle of 17.6°,and its screw pitch equals to the diameter of the screw. Each segment has different functions.

压缩段

The feeding section refers to a distance extending from the plastics inlet to forward,and it is about 4 – 8 times the length of the screw diameter. In this section,the plastics is still solid. The main function of it is to grab material from the hopper and transfer it to the compression section and to heat the material at the same time. Because of the low density of the material,the spiral groove is deep, and the screw in the feed section(h_1) is 0. 10 – 0. 15 times the length of the screw diameter, in addition, in order to create the best transport condition for the plastics,it requires to reduce the friction between material and the screw and increase the tangential friction between the material and the sleeve. In this case, the surface of the sleeve which is in contact with the plastics should open longitudinal groove (see Fig. 2 – 5), and increase the smooth extent of screw surface and cool the center of the screw with water.

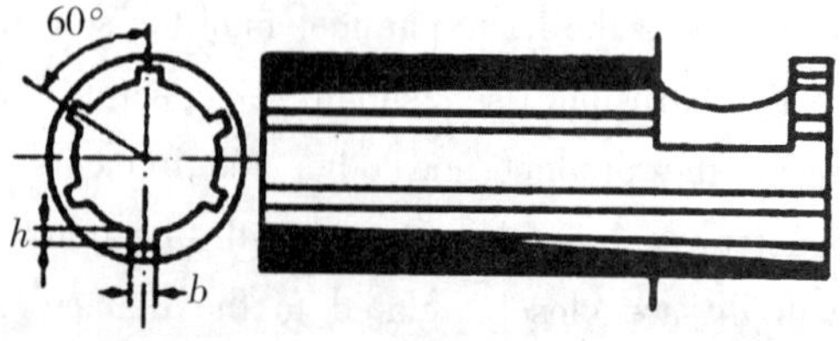

Fig. 2 – 6 The cylinder with inner surface opening longitudinal groove

Compression section (transitional section) is the middle section of screw. Apart from heating and forwarding, plastics is gradually compacted and softened from granular solids to continuous melts as well as discharges entrained air to the feeding section. Accordingly, the depth of spiral groove is gradually reduced to the screw spiral groove depth(h) in the metering section, which is not only conducive to the quality of the products, but also the warming and the melting of the materials. Typically, the ratio between the volume of a spiral groove in feeding section and the volume of a spiral groove in metering section is called screw compression ratio. For this constant pitch screw, the compression ratio can be expressed by(h_1/h_2), whose value is 2 - 4. Generally, it depends on the types of plastics, the aggregation state when it is fed and the shape of extrusion products.

压缩比

聚集状态

Metering section (homogenization section) is the last section of the screw, it is of 6 to 10 times of the length of the screw diameter. It is called metering section, because the function of it is to plasticize melt evenly and extrude material flow from the flow channel of the nose and mouth die quantitatively and in constant pressure. The depth of this spiral groove is relatively shallow, with the depth of 0.02 to 0.06 times of the screw diameter.

计量段/均化段

3. Helix Angle of the Screw and the Screw Flight Width(e)

The size of helix angle is relevant to the shape of the materials. Different material shapes require different helix angles of the feed zone. It can be proved that a spiral angle of 30° is suitable for fine powder plastics both in theory and experiment; a spiral angle of 15° is suitable for box materials; a spiral angle of 17.6° is suitable for ball and columnar materials. In the metering section, according to the formula derivation, a helix angle of 30° produces a maximum yield. However, considering the screw manufacture, it is easy to process when the pitch is equal to the diameter and the helix angle is 17.6°. Besides, it has slight effect on the yield, and the spiral direction of screw is generally right.

Screw flight is generally 0.08 - 0.12 times of the width of the screw diameter, but the bottom of the spiral groove is wider and its root uses arc transition.

圆弧

4. Shape of the Screw Head

The screw head is usually a blunt tip cone [shown in Fig. 2 - 7(a), (b)] to avoid decomposition of the materials because of long stagnant in the screw head. If it is a axial displacement screw, the screw head can also play the role of pressure adjustment. The screw head can also be torpedo-shaped [shown in Fig. 2 - 7

鱼雷头/平准头

(c)], and called torpedo head or leveling head. The gap between leveling head and the sleeve is usually smaller than the spiral groove depth, whose surface can be opened to trench or rolled into special patterns. It benefits the mixing and heating of plastics and it is conducive to increase the pressure of the feeding stream and to eliminate pulsation phenomenon, commonly used to squeeze plastics of a large extrusion viscosity and poor thermal conductivity or plastics with an apparent melting point.

导热性

(a) (b) (c)

Fig. 2 – 7 Common traits of the screw head

单螺杆挤出原理

2.3 Principle of a Single Screw Extruder

It should ensure that the plastics fed into extruder in a certain speed can be completely melted to meet quality requirements in order to guarantee the quality and yield of the extruded products. That is to say, the delivery rate of the melt should be equal to the melting rate of the materials. In addition, it is necessary to guarantee the mass flow rate in a random cross section to keep unchanged and be equal to the yield. If not, it will cause the fluctuation of the output and temperature. Therefore, only theoretically making clear of the mechanism of solids conveying, melting and melt conveying in extruder firstly can we develop reasonable process conditions according to the performance of plastics and the screw geometric characteristics.

机理

固体输送

2.3.1 Solids Conveying

Currently, the simplest way to derive solid transport theory is based on the friction force static balance between solids. There are some assumptions when deriving: ①Material is closely contacted with the screw groove as well as all edges of the sleeve wall, forming a solid plug or bed and shifting constantly. ②The effect of the gap between screw flight and sleeve is omitted and material gravity and density are changed. ③Spiral groove depth is constant, but the pressure is only a function of the groove length and the friction coefficient is independent of

螺棱

pressure. ④ Solid materials in screw groove move just like the way of elastic solids plugs. The movement of the solid plug is subject to the friction between the screw around the solid and the surface of the sleeve. Only when the friction force between the material and the screw is smaller than that between the material and the sleeve can it move axially. Otherwise, the material will be rotated together with the screw.

The main task of the solid conveying section is to transport plastics whose melting rate matches the requirement of next section, thus, its research object is solid conveying rate (Q_s).

固体输送率

As shown in Fig. 2 – 8, in order to get Q_s, firstly, we need to know that the velocity of solid plug in the axial V_{pL}, and then calculate cross-sectional area of the channel, multiply them and finally we can figure out the result. That is:

$$Q_s = V_{pL} \int_{R_s}^{R_b} \left(2\pi R - \frac{ie}{\sin\theta a}\right) dR \qquad (2-1)$$

In the formula, R_s and R_b are the radius of the spiral groove's bottom and top respectively; e is the screw flight width; θ is the average helix angle; i is the number of threaded head.

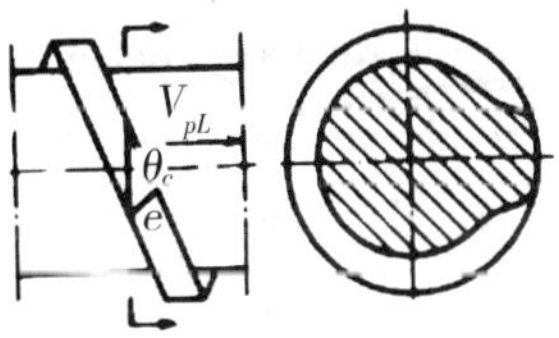

Fig. 2 – 8 Screw-sectional view

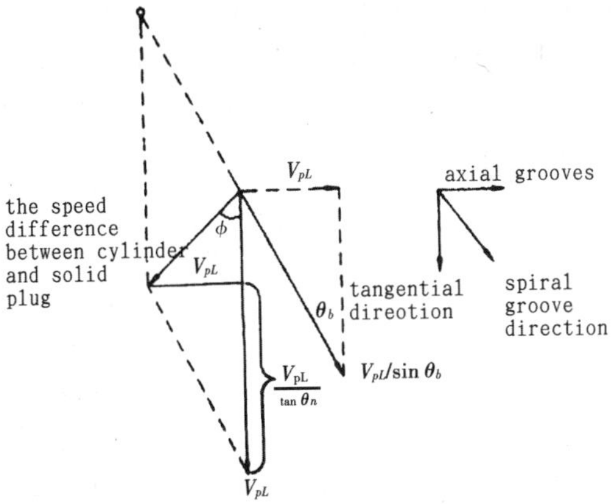

Fig. 2 – 9 The speed difference velocity vector between the sleeve and solid plug

Assumed that the screw would stay stationary and the sleeve would move relatively to the screw, its speed is $V_b = \pi D_b N$. Since the friction force between the solid plug and the screw is smaller than that between the solid plug and the sleeve, the solid plug will move along the direction of the spiral groove, and its speed is $V_{pZ} = V_{pL}/\sin\theta_b$; its tangential velocity is $V_{P\theta} = V_{pL}/tan\theta_b$. According to the synthesis of rate, as long as the angle ϕ is known, V_{PL} is solvable. Obtained from Fig. 2 - 9:

切向速度
合成

$$\tan\phi = V_{pL}/(V_b - V_{pL}/\tan\theta_b) \qquad (2-2)$$

Angle ϕ is called shift angle ($0 < \phi < 90°$). Physically speaking, the angle ϕ is located in the direction of the solid plug where observers can watch the sleeve move. Its value can be calculated based on the force acted on the solid plug and torque balance . Due to the complex result, it is omitted.

力矩平衡

On substituting V_{pL} into the formula (2 - 1), we have:

$$Q_s = \pi^2 D_b N H_f (D_b - H_f)(\tan\phi\tan\theta_b)/(\tan\phi + \tan\theta_b) \qquad (2-3)$$

In the formula, N is the screw revolution speed; θ_b is the helix angle of the sleeve at the surface; D_b is the outer diameter of the screw.

From the formula (2 - 3), the solids conveying rate is not only proportional to both $D_b N H_f (D_b - H_f)$, but also to tangent functions [$\tan\phi\tan\theta_b/(tan\phi + tan\theta_b)$]. For the latter, because the displacement angle is related to the geometric parameters of the screw and the sleeve, and the friction coefficient (f, f_b) is also connected with the the pressure drop of solids conveying section, in order to calculate briefly, the pressure drop of conveying end is omitted and $f_s[\tan\phi\tan\theta_b/(\tan\phi + \tan\theta_b)] = f_b$ is mapped to helix angle θ (as shown in Fig. 2 - 10). We can see from the figure that tangent function will appear the maximum value in a specific helix angle if f_s has been determined. On the other hand, the optimal helix angle increases while the coefficient of friction decreases. According to the known experimental data, f_s of most plastics is in the range of 0.25 to 0.50, thus the best helix angle is from 17° to 20°.

正切函数
几何参数
摩擦系数

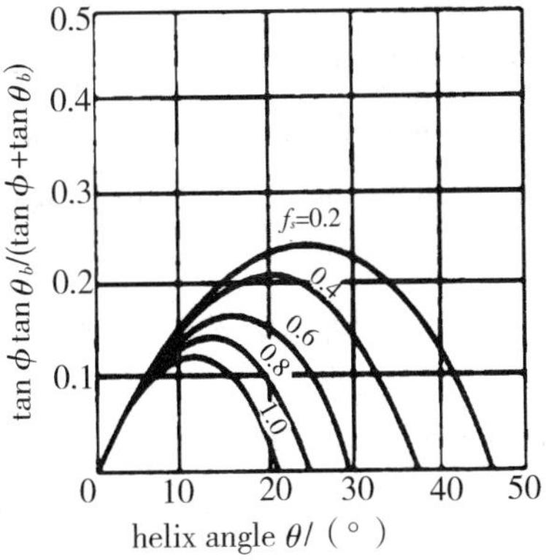

Fig. 2 – 10 Relationship between the tangent function and helix angle

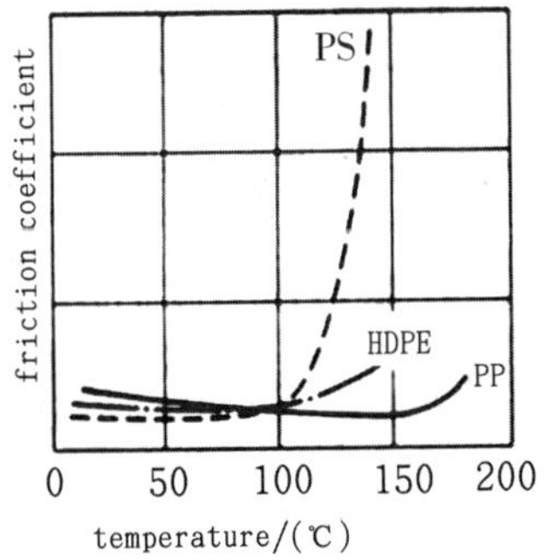

Fig. 2 – 11 Steel material's temperature dependence of the friction coefficient

In order to obtain maximum solids conveying rate, measures can be taken from the structure of the extruder and extruding process. Taking consideration of the extruder structure, it is good to increase the grooves depth, but it will be subject to a screw torque limit. Besides, it is advantageous to reduce the friction coefficient between the plastic and the screw (f_s), which is easy to improve the surface finish of the screw (lowering the surface of roughness of screw machining). Furthermore, it can increase the solids conveying rate by increasing the friction coefficient between the plastic and the sleeve. In this case, it seems that the sleeve surface should be rough, but this will cause the material's stagnation and even decomposition, so the inner surface of the sleeve should be as smooth as possible. There are some effective ways to improve sleeve friction coefficient. ① Open longitudinal grooves in the sleeve. ② Adopt conical slotted sleeve (see Fig. 2 – 6). ③ The helix angle is adopted as 17°41′ instead of its optimum value due to taking

螺杆扭矩

manufacturing convenience into account.

Taking consideration of the extrusion process, the key is to control the temperature of the feeding section of the sleeve and the screw because the friction coefficient varies with temperature. The relationship between the friction coefficient of some plastics to steel and temperature is shown in Fig. 2 – 11. After determining

移动角 the geometric parameters of the screw, the movement angle is only related to the friction coefficient.

If the friction between the material and the screw is so large that the material clings to the screw, the extrusion rate Q_s and the moving speed are both zero because $F = 0$. In this time, the materials cannot move forward, which is also called "do not feed". If the friction between the material and the screw is very small, even negligible, and the friction to the sleeve is great, the materials will move forward within a very high speed, that is $\phi = 90°$. If there is open longitudinal groove in the sleeve which forces the material to move forward along $\phi = 90°$ direction, it is the maximum limit of the solids conveying rate theoretically. It ranges from 0 to 90 degree in general situation. During the extrusion process, if the friction between the material and the screw and the sleeve

恒定值 is not a constant value, it is bound to cause changes of the movement angle and

波动 the fluctuation of yield.

2.3.2 Solid Melting

固体融化

Plastics in extruder is heated by external heat and internal heat (the friction within materials and the friction between the materials and the metal) to raise temperature. Thus, solid plastics will be gradually melted and completely become the melt in the end. So, there must be a region where solid and melt are

熔化区/相变区 coexistent, that is the melting zone or the phase transition region. Since in this area there is two-phase coexistence, it brings many difficulties to study, and we haven't obtained a reasonable theory until 1966. The theoretical derivation is very cumbersome, so we only give a brief introduction here.

1. Melting Process

The materials fed by conveying section are in contact with the heated tube surface after entering the melting zone, where the advanced process is. Melting

熔体膜 begins from the contact portion, leaving a layer of melt film on the surface of the sleeve. If the thickness of the melt film exceeds the gap between screw flight and the cylinder, it will be scraped off by the rotated screw flight, forcing to accumulate it at the front side of screw flight to form a melt pool ,while the rear side of the screw flight will form a solid bed . It is shown in Fig. 2 – 12. Thus, in the process of the grooves moving forward, the width of the solid bed is gradually reduced until it totally disappeared, i. e. it is completely melted (this process has been experimentally confirmed). The total length starts from the melting to the solid bed width dropping to zero, which is known as the length of the melting zone. 熔体池 固体床

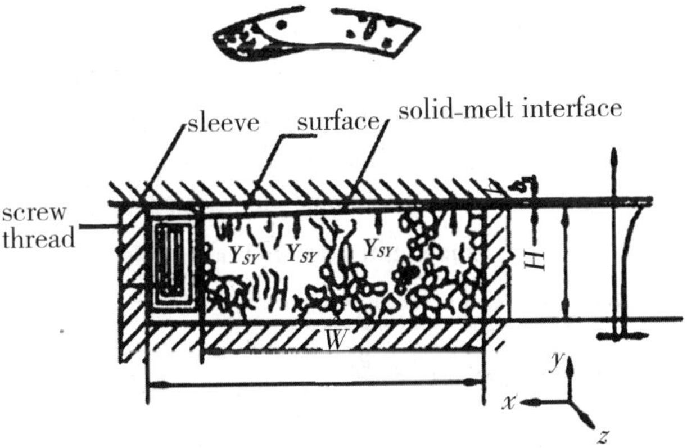

Fig. 2 – 12 Spiral groove cross-section

Solids melting after formation of the melt film occurs at the interface of the melt film and the solid bed. Part of the needed heat comes from the sleeve heaters, and the rest comes from shearing function of the screw and the sleeve to the melt film. 料筒加热器

2. Melting Rate

The main purpose of studying melting theory is to make predictions about the quantity of unmelted material in any point of the spiral groove, the required screw length to melt all materials, and dependent relationships among these two variables on the material properties, the geometry of the screw and operating conditions. In order to explain this question briefly and clearly, we take Tadmor melting model as an example for analysis. In order to establish the mathematical model of the melting process, we take some basic assumptions particularly. ① Melting process is steady. ② Material in screw groove is compacted into a continuous and uniform solid bed (plug) moving at a constant speed along the spiral groove. ③ Spiral groove is rectangular and the gap between the screw flight and the sleeve can be negligible. ④ Interface boundary is obvious, i. e. the plastics has an obvious melting point. ⑤ Solids melting only happens in the solid bed between the sleeve and

热传导

牛顿流体

the solid / melt film interface. ⑥ Heat conduction and flow are one-dimensional, i. e. the temperature and velocity just have the functions of distance from the interface with X and Z direction being not taken into account. ⑦ Melt is Newtonian fluid.

Assumptions above indicate that the melting of the material in the screw groove happens on the melt-solid interface, which is subject to the interface, so that the heat consumed for the material melting is different between entering and leaving. In this case, it is necessary to know the temperature distribution of the melt film and the solid, and the input and output of the heat.

If the screw is fixed, the sleeve moves with the velocity of V_b. The component velocity in the direction of the groove is V_{bz}; the velocity entering into the interface is V_{sy} and the relative velocity of the solid bed is V_j; the thickness of the melt film is δ; the temperature which is away from the interface of the solid bed is t_s; the melting point of the materials is T_m; the temperature of the sleeve surface is T_b; the melt viscosity is μ. The temperature distribution in the melt film is obtained from the energy equation and the boundary conditions (Fig. 2 – 13).

$$\frac{T - T_m}{T - T_b} = \frac{\mu V_j^2}{2k_m(T_b - T_m)} \times \frac{y}{\delta}\left(1 - \frac{y}{\delta}\right) + \frac{y}{\delta} \tag{2-4}$$

热导率

勃林克曼数

In the formula, y is the distance away from the interface; k_m is thermal conductivity; the dimensionless group $\mu V_j^2/2k_m(T_b - T_m)$ is usually known as Bolinkeman number, which represents the ratio of the heat produced by the sleeve when the difference of the heat and the temperature generated by shearing that is $(T_b - T_m)$. If Bolinkeman number is larger than 2, the temperature in some position between the sleeve and the interface may be higher than T_b, because shearing generates much more heat.

The heat which comes from the melt film to enter into the unit interface is:

$$-(qy)_{y=0} = k_m\left(\frac{dT}{dy}\right)_{y=0} = \frac{k_m}{\delta}(T_b - T_m) + \frac{\mu V_j^2}{2\delta} \tag{2-5}$$

The temperature distribution within the solid bed can be inferred in the boundary conditions when $y = 0, T = T_m$ and $y \to -\infty, T \to T_s$:

$$\frac{T - T_s}{T_m - T_s} = \exp\left(\frac{V_{sy}}{\alpha_s} y\right) \tag{2-6}$$

热扩散系数

比热容

In the formula, $\alpha_s = k_s/\rho_s C_s$ is the thermal diffusivity; k_s, ρ_s and C_s are thermal conductivity, density and specific heat capacity of the solid bed. Formula (2 – 6) describing the solid bed temperature is the initial temperature which decreases exponentially from the melting point T_m to the starting temperature of the solid bed.

Heat which transmits from the melt film to solid in unit interface is:

$$-(qy)_{y=0} = k_s\left(\frac{dT}{dy}\right)_{y=0} = \rho_s C_s V_{sy}(T_m - T_s) \qquad (2-7)$$

Temperature distribution curve from the sleeve to the solid bed is shown in Fig. 2 – 13.

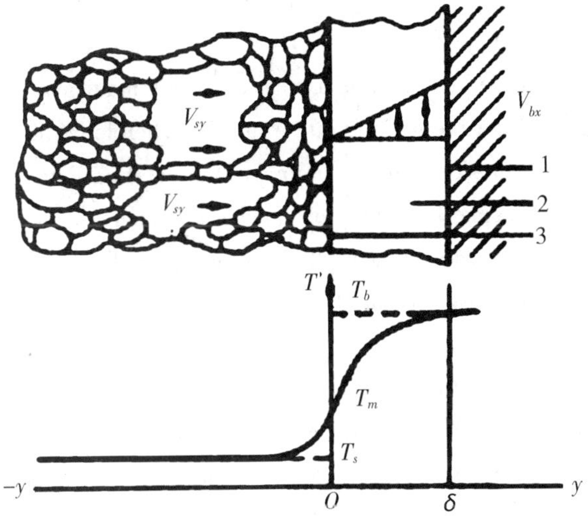

Fig. 2 – 13 The temperature distribution in melt film and solid bed
(1—sleeve surface; 2—melt film; 3—interface)

Consolidating formula (2 – 6) and formula (2 – 7), we can know that the difference between the input and output of heat of unit interface is the consumed heat when melting the materials:

$$\left[\frac{km}{\delta}(T_b - T_m) + \frac{\mu V_j^2}{2\delta}\right] - [\rho_s C_s V_{sy}(T_m - T_b) = V_{sy}\rho_s \lambda] \qquad (2-8)$$

In the formula λ is the melting heat of the plastics. 熔化热

Considering the materials' balance, the amount of solids entering the melt film from interface should be equal to the outflow of the melt. There is:

$$\omega \equiv V_{sy}\rho_s X = \frac{V_{bx}}{2}\rho_m \delta \qquad (2-9)$$

In the formula, ω is defined as melting rate of a unit of groove length. After solving V_{sy} from the equation (2 – 8) and substituting it into the formula (2 – 9), the thickness of the melt film δ and the melting rate ω can be expressed by the width of the solid bed X:

$$\delta = \left\{\frac{[2k_m(T_b - T_m) + \mu V_j^2]X}{V_{bx}\rho_m[C_s(T_m - T_s) + \lambda]}\right\}^{1/2} \qquad (2-10)$$

$$\omega = \left\{ \frac{V_{bx}\rho_m [k_m (T_b - T_m) + \frac{\mu}{2} V_j^2] X}{2[C_s (T_m - T_s) + \lambda]} \right\}^{1/2} = \phi X^{1/2} \tag{2-11}$$

$$\phi = \left\{ \frac{V_{bx}\rho_m \left[km (T_b - T_m) + \frac{\mu}{2} V_j^2 \right]}{2[C_s (T_m - T_s) + \lambda]} \right\}^{1/2} \tag{2-12}$$

变量群

Variable group defined by φ is a measure of the melt rate, i. e. the larger the value of ϕ is, the higher the melting rate would be. Therefore, the following aspects can be taken into consideration to improve the melting rate of the material: ① Without compromising the quality of the sleeve to increase the temperature of it and increase external heating supply, there is an optimal value because when the temperature rises, the viscosity of the materials decreases and the friction heat relatively reduces. ② Preheat the materials to increase the materials' temperature so as to reduce the amount of heat needed for melting. ③ Increase the screw speed for certain materials, because improving screw speed can increase the friction heat of the materials, but if improving the flow at the same time, the increase of the speed would be meaningless when the requiring heat of the increased flow melt is greater than the increase of the friction heat. Otherwise it will be advisable to increase the melting rate.

黏度

2.3.3 Melt Conveying

The theory of the melt conveying in the metering section is one of the earliest and the most complete theories in the single-screw extruder researches. It starts from taking isothermal Newtonian fluid between two infinite parallel plates as objects and deals with the issue in extremely simplified situation. Later, it extends to non-Newtonian fluid. As time goes by, the theoretical research gets closer to the fact, but it is much more complicated. Here, we will only discuss the simplest one. It should be noted that the formula calculation led by the simplest theory has its accuracy in a certain degree with simple calculation.

流动方程

1. Simplify Fluid Equation

To facilitate the derivation easily, the spiral groove is expanded and positioned in the plane, and the sleeve is seen as a mobile tablet on the spiral groove and moved with at a constant speed (see Fig. 2 - 14). Assuming that: ① The melt is incompressible Newtonian fluid, whose viscosity (mu) is independent from temperature. ② Current is a stable laminar flow which is fully developed. ③ Fluid does not slide in the wall surface. ④ Spiral groove is

层流

rectangular and the depth of the groove(H_m) is much smaller than the width(W) of the spiral groove ,i. e. $H_m \leq W$. ⑤ The influence of the inertial force, gravity and other factors are not considered. ⑥ Physical properties such as melt density remain unchanged. ⑦The depth of the spiral groove and the pressure gradient (dp / dz) are both constant.

物理性质
压力梯度

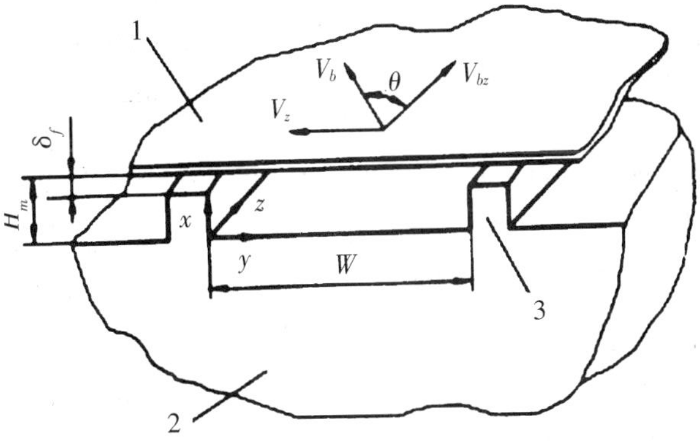

Fig. 2 – 14 Expanded view of the spiral groove
(1—sleeve;2—screw root;3—screw wing)

There are four flows for the melting in the screw metering section, containing positive flow (drag flow), counter flow (pressure flow), cross-flow and leak (vent) flow (See Fig. 2 – 15). Positive flow (Q_d) is generated when the sleeve moves in the direction of the groove. Counter flow(Q_p), whose direction is opposite to that of positive flow, is generated by the pressure gradient of the feeding stream. Cross-flow(Q_t) is the flow generated when the materials move along the z-axis. In order to ensure the continuity of cross flow, the materials also move along the y-axis, thus forming a circulation affecting its mixing and heat-transferring, but not the flow. Leak flow(Q_L) which is also caused by the pressure gradient is the flow when materials move from a gap between the spiral rim and the sleeve(δ_f) along the direction of the screw axially to the hopper.

正流/逆流/横流/漏(泄)流

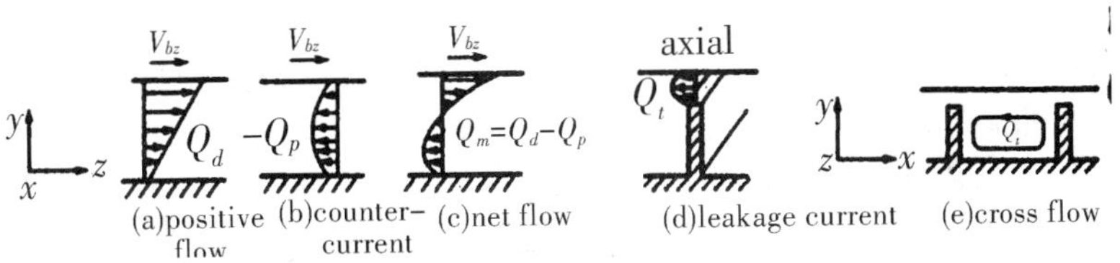

Fig. 2 – 15 The current forms

If the speed of materials along the spiral groove is V_z, the loss of leakage flow current and the influence of the cross-flow are not considered. It can be obtained by combining the momentum equation with Newton's law of viscosity:

动量方程/牛顿黏性定律

$$\frac{d^2V_z}{dy^2}=\frac{1}{\eta}\times\frac{dp}{dz} \tag{2-13}$$

Integrating the above formula and adopting appropriate boundary conditions, then we have:

$$V_z=yV_{bz}/H_m-y(H_m-y)(dp/dz)/2\mu \tag{2-14}$$

特性曲线
速度分布

The first term of the above formula on the right side represents velocity distribution of positive flow; the second term represents velocity distribution of counter flow; the sum is the velocity distribution of the melt conveying rate of net flow in z direction. As shown in Fig. 2-15.

净流

The melt transfer rate (net flow) is derived by integrating the velocity distribution along the spiral groove and the cross-sectional scale of the spiral groove:

常数

$$Q_m=\int_b^{H_m}\int_0^w v_z dydx=\frac{V_{bz}WH_m}{2}-\frac{WH_m^3}{12\mu}\left(\frac{dp}{dz}\right)=\frac{1}{2}V_{bz}WH_m-\frac{WH_m^3\Delta p}{12\mu z} \tag{2-15}$$

负斜率
压力降

In the formula, z is the length along the spiral groove; Δp is the pressure drop.

It can be drawn from formula (2-15): ①When $\Delta p=0$, $Q_{max}=\frac{1}{2}V_{bz}WH_m$. ②When $Q_m=0$, $\Delta p_{max}=6\mu V_{bxz}/H_m^2$, i. e. from the mouth die, there is no extruded materials to generate the greatest pressure drop. ③ Normal extrusion operation is usually carried out between the previous two extremes.

In addition, it can be seen that positive flow is proportional to the depth of screw groove (H_m), while the counter flow is proportional to its cubic. Therefore, when the pressure is lower, extrusion amount of shallow grooves screw will be lower than that of deep grooves screw, while it is contrary when the pressure increases to a certain degree. It indicates that the shallow grooves screw is not very sensitive to pressure and can extrude uniform products in the case of pressure fluctuation. However, the spiral groove should not be too shallow, or it will be easy to burn plastics.

塑料熔体

Make use of the screw geometric relationship. The simplified flow equations can be written as:

口模形状

$$Q_m=\frac{\pi^2D_b^2NH_m\sin\theta_b\cos\theta_b}{2}-\frac{\pi D_bH_m^3\sin^2\theta_b\Delta p}{12\mu L} \tag{2-16}$$

In the formula, N is the rotational speed of the screw; Δp is the pressure drop of the material flow in the metering section; L is the length of the metering section. 转速

2. Characteristic Curves of Screw and Die 特性曲线

Simplified, formula (2 – 15) can be rewritten as the following formula:

$$Q_m = AN - B\frac{\Delta p}{\eta} \quad (2-17)$$

In the formula, A and B are only associated with the size of the screw structure, producing Newtonian fluid to specified extruder under isothermal condition. Except for Q_m and Δp, other symbols in formula (2 – 17) are constants. 常数 This formula (2 – 17) is the equation of a straight line. If it is plotted on the graph of the $Q_m - \Delta p$, it can obtain a series of parallel straight lines with a negative slope, which are often called the screw characteristic curves (see Fig. 2 – 16). 负斜率

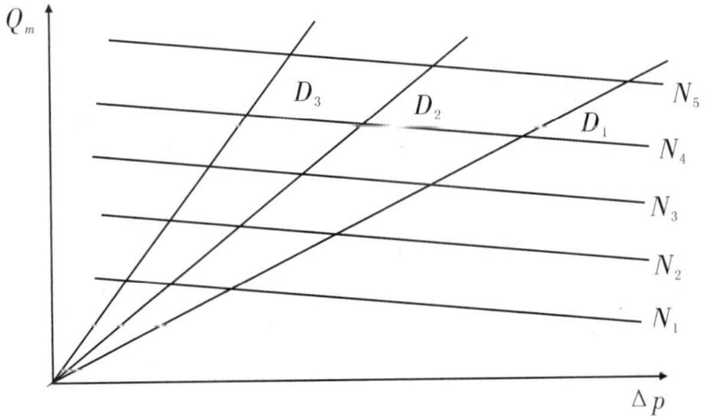

Fig. 2 – 16 Characteristic curves of the screw and die

Similarly, that Newtonian plastic melt going through the die equation can be abbreviated as: 塑料熔体

$$Q_m = K\frac{\Delta p}{\eta} \quad (2-18)$$

In the formula, K is a constant, relevant to the die geometry. Δp is the pressure drop when the plastics goes through the die. Because the formation of pressure in metering section aims to allow the plastic melt to overcome its resistance when flowing through the die, since in most cases, the pressure of material flow in metering section is equal to the pressure of the die at the discharge. Thus, Δp in formula (2 – 18) is equal to the Δp in formula (2 – 17). Plotting formula (2 – 18) in the same coordinate, it can obtain another group of straight lines as shown in Fig. 2 – 16 (D_1, D_2, D_3, etc., different straight lines are 口模形状

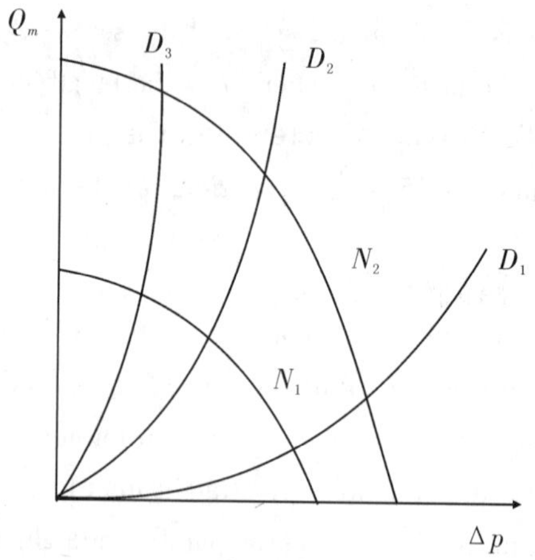

Fig. 2 – 17 Characteristic curves of plastic melt screw and die

represented with diffe rent dies, that is, the value of η is different). This kind of straight lines is called die characteristic curves.

口模特性曲线

操作点

Intersection of the two sets of straight lines in Fig. 2 – 16 is the operating point. By making use of this graph, extrusion amounts of specific extruders matched with different dies can be obtained. It is extremely easy to use because a straight line can be decided only by two points. Make simultaneous equations (2 – 17) and (2 – 18) and eliminate Δp, then we get:

$$Q_m = \left(\frac{AK}{K + B}\right)N \tag{2-19}$$

It is shown in (2 – 19) that extrusion amount of the extruder (with extrusion die) is only related to the rotational speed of the screw and the size and structure of the screw and die, but it has nothing to do with the plastic viscosity.

假塑性熔体

抛物线

In the isothermal case, viscosity of pseudo plastic melt is not a constant, and its screw and die characteristic curves are no longer straight lines but parabolic curves (see Fig. 2 – 17). It has been already proved by experimental measurements.

It should be noted that the characteristic curves of the screw and the die only reflect the relationship between the flow rate and the die pressure, not the relationship between the quality of extrudate and other conditions. Therefore, under the premise of ensuring product quality in the production, production can be increased by adjusting the pressure and other conditions (such as rotational speed, sleeve temperature, etc.).

Chapter 3
Injection Molding

注射成型

3.1 Outline

概述

Injection molding (also known as injection moulding or injection) is an important method of molding plastic products. Almost all thermoplastics and plenty of thermoset plastics can use this method of molding. Using injection molding can produce molded articles of various shapes, sizes, precisions to meet various requirements. Injection molding products account for 20% – 30% of total plastic products. Especially when plastic materials appear as engineering structure materials, the application of the injection-molded article has extended from the civil life to all fields of the national economy, and will gradually replace the conventional metal and products of non-metallic meterials, including various industrial parts, instrumentation parts and structures, housing materials and so on. In the cutting-edge of technology, it is also indispensable.

成型塑料制品
热固性塑料
工程结构材料

The injection molding process starts from feeding the granular plastic or powdery plastics from the hopper of the injection machine to the heated sleeve (See Fig. 3 – 1). After being heated and melted into fluid completely, under the push of the plunger or the screw, it goes through the front of the sleeve nozzle into a closed mold. Melt which is filled with the mold, solidified by cooling (thermoplastics) or heating (thermoset plastics) under pressure, can maintain the shape of sample which the injection mold cavity confers. Opening the mold to obtain the article means a molding cycle is completed. After that, it is the repetition of the same production process.

注射模塑过程
喷嘴
模塑周期

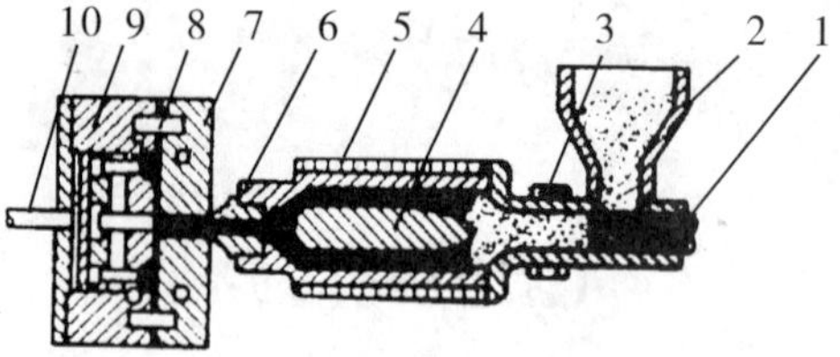

Fig. 3 – 1 The profile of injection machine and plastic film

(1—plunger; 2—hopper; 3—cooling jacket; 4—shunt shuttle; 5—heater; 6—nozzle; 7—fixed template; 8—product; 9—active template; 10—ejector)

An injection molding cycle ranges from a few seconds to several minutes. The length of the time depends on the types of articles in size, shape, thickness and the sorts of the injection machines, plastic varieties, process conditions and other factors. The weight of each product ranges from lower than 1gram to tens of kilograms, depending on the specifications of injection molding machine and the requirements of products.

工艺条件

Injection molding has a series of advantages including short molding cycle, forming articles of complicated shape, accurate size and inserting in one time. Besides, it has strong adaptability to various plastics and it owns high efficiency and is easy to implement fully automatic production. It is a kind of technology which is fully advanced, developed and economical. It will be developed towards the direction of high-speed and automation.

注射机

Injection is achieved through the injection machine. There are many types of injection molding machines, no matter what type of the machine is, the basic roles of it are to: ① heat the plastics to reach the molten state; ② apply a high pressure on the molten plastics to make them eject to fill the mold cavity . In order to better accomplish these two fundamental functions, the structure of the injection molding machine has gone through continuous improvement and development. Firstly what appeared is the plunger-type injection machine (Fig. 3 – 1) with simple structure, which achieves two fundamental roles of plasticization and injection by means of a sleeve and a piston, but achieving the control of the temperature and pressure is more difficult.

型腔

单螺杆定位预塑注射机

Then is the positioning of the single screw pre-plasticizing plunger-type injection machine (see Fig. 3 – 2), formed by the convergence of the pre-plastic sleeve and injection sleeve. The plastics is heated and plasticized in the pre-plastic sleeve and then squeezed into the injection sleeve. Then it is injected into the mold cavity by the plunger under high pressure. This kind of injection machine feeds a large amount of materials. Its plasticizing effect has been significantly improved and it owns more stable injection pressure and speed, but its operation and structure are more complex. Thus it is not widely applied. Finally is the mobile screw injection machine which is composed by a single screw and a sleeve (Fig. 3 - 3). Added plastics is heated and plasticized relying on the rotation of the screw in the sleeve, and continues to be pushed to the sleeve distal end, which is close to the nozzle, so the screw moves backward slowly while rotating. When retreating to a predetermined position, screw stops rotating. In this case, the screw receives high pressure passed by the hydraulic cylinder plunger and axial displacement happens, pushing the molten plastics accumulated in the sleeve end through the nozzle into the mold at a high speed. The efficiency of mobile screw injection machine is almost the same as pre-plasticizing plunger-type injection machine with simple structure and easy manufacturing. Compared with a plunger-type injection molding machine, it can make plastics get a good mixing and plasticize in the sleeve, not only improving the quality of molding plastics but also expanding the range of the plastic injection molding and the volume of injection. Therefore, the mobile screw injection machine has overwhelming advantages in the development of injection machine.

柱塞式注射机

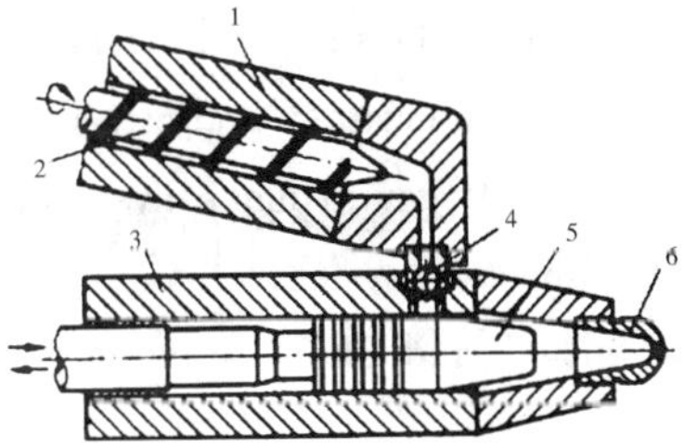

Fig. 3 - 2 Screw pre-plasticizing plunger-type injection machine

(1—pre-plasticizing cylinder; 2—extrusion screw; 3—injection cylinder; 4—ball-type non-return nozzle; 5—injection plunger; 6—nozzle)

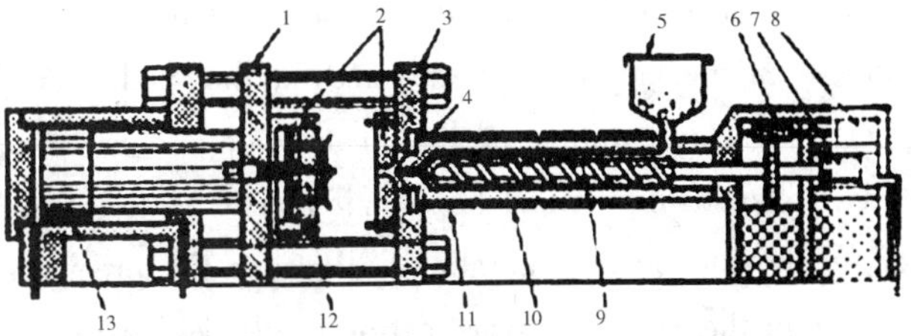

Fig. 3 - 3 Mobile screw injection machine

[1—moving plate; 2—injection mold; 3—fixed template; 4—nozzle; 5—hopper; 6—screw drive gear; 7—injection cylinder; 8—hydraulic pump; 9—screw; 10—heating barrel; 11—heater; 12—ejector rod (pin); 13—clamping cylinder]

Mobile screw injection machine is widely used in factories nowadays, but there is still a small amount of application of a plunger-type injection machine. It is generally used to produce small parts whose weight is under 60 g. But when we refer to the production of heat-sensitive plastic, the illiquid plastic, medium and large-scale injection molding machine, the mobile screw injection machine takes the burden. Therefore, this section will introduce these two devices and their techniques of molding thermoplastic.

注射模塑设备

3.2 Injection Molding Apparatus

锁模系统/塑模

Mobile screw injection machine and plunger-type injection molding machine are both composed of the injection system, clamping system and plastic mold, which will be described below.

注射系统

3.2.1 Injection System

分流梭

It is the most important part of the injection machine, whose role is to achieve the plastic homogenization and plasticization, and at very high pressure and speed, it pushes through the screw or plunger to make even and plasticized plastics be injected into the mold. The injection system is comprised of a feeding device, a material sleeve, a screw (piston and shunt spindle for plunger type injection machine), a nozzle, etc.

1. Feeding Device

自动上料装置

The feeding device of a small injection molding machine is usually a conical hopper which is connected to a sleeve. Hopper capacity is about the feeding amount of production of 1 to 2 hours. If capacity is too large, the plastic will reabsorb moisture from the air, which is adverse for the quality of the products. Only the hopper with configuration of the heating apparatus can appropriately increase capacity. When using conical hopper, if plastic particles are uneven, the vibration generated by the device operation will cause the precipitation of small particles or powder in the hopper, thus affecting the bulk density of the materials and causing uneven feeding before and after. When the hopper is used for the plunger injection molding machine, it should be configured with constant quantity or constant volume feeding device. The hopper of large injection machine is basically tapered, with other automatic feeding device.

2. Sleeve

The sleeve is taken as the container for the plastic heating and pressing, thus it is required that its sleeve is able to withstand pressure, having heat, fatigue and corrosion resistance and good heat conductivity. The volume of plunger injection molding machine sleeve is 4 to 8 times of its maximum injection amount. If the volume is too large, the plastics will be heated for a long time in the high temperature within the sleeve, causing the plastics decomposition and discoloration , affecting product quality, or even interrupting the production; however, if the volume is too small, it will cause uneven plasticization. Since screw injection machine has the screw to stir plastics in the sleeve, the material layer is relatively thin and it owns high heat-transferring efficiency and the uniformly plasticization. Thus, the volume of a sleeve is 2 to 3 times of its maximum injection amount. The outside of the sleeve is equipped with heating device; usually the sleeve is divided into 2 to 6 heating zones to enable segmented heating and control. Temperature of the end close to the hopper is lower than the temperature of the end close to the nozzle. Sleeve temperatures of each section are precisely displayed by the thermocouple and controlled by temperature controlled instrument. The corner of the inner wall of the sleeve should be made streamlined to prevent the deposit of material from affecting the quality of the products. The machinery of each part in the sleeve should keep precise cooperation.

耐疲劳/耐腐蚀

分解

变色

3. Plunger and Shunt Spindle

The plunger and shunt spindle are both important components of the sleeve in plunger injection machine. The role of the plunger is to pass the pressure of injection cylinder to the plastics and make melt inject into the mold. The plunger is a solid metal column of high hard surface, the diameter of which is usually from 20 to 100 mm. The proportion of the cross-sectional area of the injection cylinder and the plunger ranges from 10 to 20. The maximum injection capacity of the injection machine in each injection is the multiply of the cross-sectional area of the plunger and the stroke of the plunger. The gap between the plunger and the sleeve should be able to stick to the role on the condition that the plunger can freely reciprocate without draining the plastics.

The split spindle is a kind of metal part which is similar to the shape of the torpedo body and located in the front of inner cavity of the sleeve distal end. Its role is to make the plastic dispersion into thin layers and evenly place in or flow through the sleeve and shunt spindle channels composed of shunt spindle, so that it

传热导程

热导率

热分解

reduces the process of lead of heat transfer, accelerates the heat-transferring and improves the quality of plasticization. The heat of plastics which is needed in the plunger-type injection machine is supplied mainly by the external sleeve heater. But the plastic flow in the sleeve is usually laminar flow, suffering lower shear rate and larger viscosity, adding that the thermal conductivity of plastics is low, so, if you want to raise the sleeve temperature gradient to increase the amount of heat transfer, achieving uniform plasticizing, it will not only extend the plasticizing time, but also make the plastics which is near to the sleeve portion be prone to thermal decomposition. Equipped with the shunt spindle, the material layer becomes thinner, which will be conducive to heating. Furthermore, the shunt spindle can also make the heat of the sleeve imported quickly by the function of positioning of the wall ribs, which is close to the sleeve, so that the shunt spindle can play the role of melting plastics from inner part, making the thin layers of plastic distributed within the channel heated by both sides and accelerating temperature fast and evenly. In addition, in the channel, due to the reduction of cross-sectional area of the material layer, the shear rate and friction heat which the melt suffers will both increase so that the viscosity will go down doubly which is beneficial to injection and heat transfer. Shunt spindle of some injection machines is also equipped with a heater, which is more conducive to the heating of the plastics.

4. Screw

The screw is an important component of the mobile screw injection machine. Its role includes plastic transport, press forming, plasticizing and pressuring. When the screw rotates in the sleeve, plastics in the hopper should be rotated into the cylinder and gradually pushed forward, compacted, exhausted and plasticized. Then the molten plastics is constantly pushed to the gap between the top of the screw and the nozzle, and the screw itself moves backward slowly due to the pressure of the melt. When the accumulation of melt reaches the volume of one injection, the screw will stop rotating. During injection, the screw passes hydraulic or mechanical force so that the melt is injected into the mold during injection. The structure of the screw is substantially the same as that of an extruder, but each has its own characteristics. ① Injection screw has axial displacement in rotation, therefore the effective length of the screw is variable. ② The aspect ratio and the compression ratio of the injection screw are relatively small. Usually L/D is 16 to 20; the compression ratio is 2 to 2.5. The injection screw only needs to conduct plastic plasticization when rotating, without requiring to provide a stable pressure because the pressure of the plastics withstands during the plasticizing is achieved by adjusting back pressure. ③ The deeper the spiral groove of injection screw is, the higher the productivity will be. ④ Because the screw of injection has axial displacement, the feed zone should be longer, which is about half the length of the screw, while the compression section and metering section account for a quarter of the length of the screw respectively. Typical injection screw is shown in Fig. 3 – 4. ⑤ In order to prevent the melt from accumulating or reflowing along the spiral groove during injection, the structure of the head of the screw should be considered, as shown in Fig. 3 – 5. The higher viscosity of the plastics often uses conical head injection screw. Adopting this screw can reduce plastics degradation. For plastics of lower viscosity, the screw head needs to be equipped with a non-return ring (as shown in Fig. 3 – 6). When it rotates, the melt advances along the groove and pushes the non-return ring forward at the same time. It goes along the clearance of the non-return ring and screw head to enter into the increasing front end of the sleeve. During injection, due to the increasing pressure of the melt in the front end of the sleeve, the non-return ring is pressed to retreat backward to the adhesion of the screw end face, thereby preventing the material from reflux.

背压

止逆环

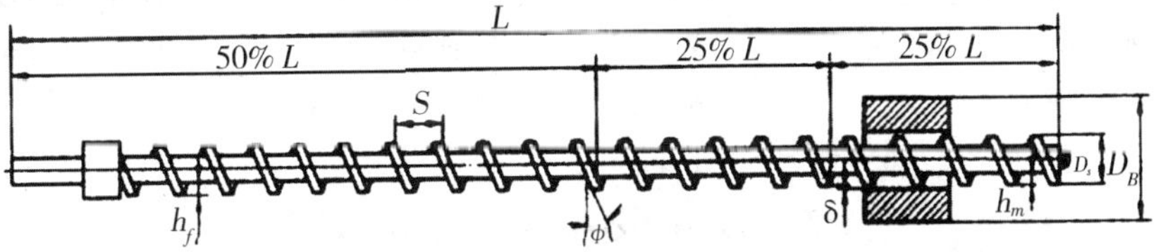

Fig. 3 – 4 Injection screw

(D_B—screw diameter; h_m—metering section spiral groove depth; D_s—screw nominal diameter; L—total length of the screw; ϕ—helix angle; δ—radial clearance; S—pitch; h_f—feeding section spiral groove depth)

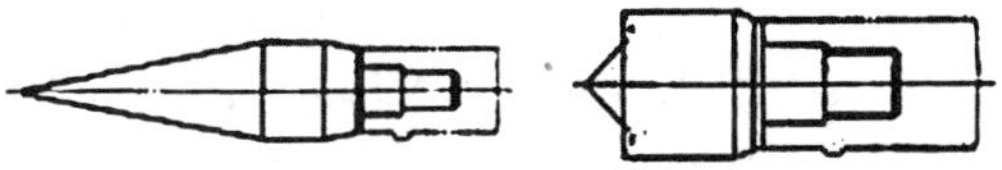

Fig. 3 – 5 Screw head

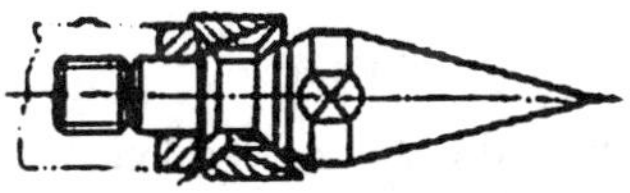

Fig. 3 – 6 Screw head with a check ring

5. Nozzle

The nozzle is the transition portion which links the sleeve to the die. During injection, the melt in the sleeve flows through the nozzle fast into the mold under the high pressure by the action of the screw or the plunger. Therefore, the structure of the nozzle, the size of the nozzle hole and the manufacturing accuracy will affect the pressure of the melt, the loss of temperature, the range of distance, the merits and shortcomings of shrinkage effect, and whether the phenomenon of "casting" happens or not. Currently there is a wide range of nozzles and they all have their own applications. Here we only discuss three kinds that are widely used.

补缩作用/"流延"现象

(1) Straight Nozzle.

直通式喷嘴

This kind of nozzle is with short tubular, as shown in Fig. 3 – 7. The loss of pressure and heat are both small when the melt flows through this nozzle and it is difficult to produce material's stagnation and decomposition, so its external will generally not get attached to heating device. However, because the nozzle body is short and the length of nozzle which goes into fixed template hole is limited, the sprue should be longer. In order to compensate this defect, increasing the length of the nozzle becomes a modified form of the straight-type nozzle, also known as the extended nozzle. This nozzle must be added to the heating device. In order to filter out the solid impurities in the melt, the nozzle may also be equipped with additional filter. These two kinds of nozzles are suitable for processing plastics of high viscosity. When processing low viscosity of plastic, there will be the phenomenon of casting.

主流道

延伸式喷嘴

滤网

Fig. 3 – 7　Straight nozzle

(2) Self-locking Nozzle.

In order to prevent extension or retraction of the melt during the injection process, it is necessary to impose temporary block on nozzle passage to use self-locking nozzle. Spring-loaded and needle-type are two main sorts of self-locking nozzles, as shown in Fig. 3 – 8. They rely on spring pressure and spool in nozzle body to achieve self-lock. During injection, the injection spool is forced to open by the resets and becomes self-lock with the role of the spring. Its advantages are obvious. Firstly, it can put an end to the casting phenomena when injecting the low viscosity plastic. Secondly, it is easy to handle and self-lock high pressure of the

自锁式喷嘴/弹簧式/针阀式

阀芯

melt and the melt is made to inject into mold. At the end of the injection, the spool effect is remarkable. However, its structure is more complex; the loss of injection pressure is huge; the range is short and its requirement of spring is demanding.

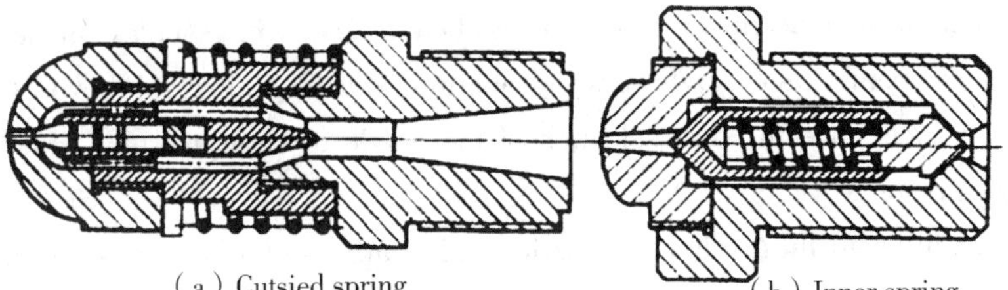

(a) Cutsied spring needle-type nozzle　(b) Inner spring needle-type nozzle

Fig. 3 - 8　Spring needle-type nozzle

(3) Leverage Needle Valve Nozzle.

杠杆针阀式喷嘴

This kind of nozzle is the same as self-locking nozzle with the implementation of the temporary opening and closing to nozzle channel during the injection process, and its structure and working principle are shown in Fig. 3 - 9. It uses an external hydraulic system via a lever to control linkage to open and close the spool. It can accurately and timely open the spool out of the need according to the manipulation of the hydraulic system. Its merits are the convenient operation, reliable self-locking, less pressure loss, accurate measurement, etc. In addition, it does not use a spring, so there is no need to take the replacement of the spring into account, but the main drawback is that the structure is relatively complicated.

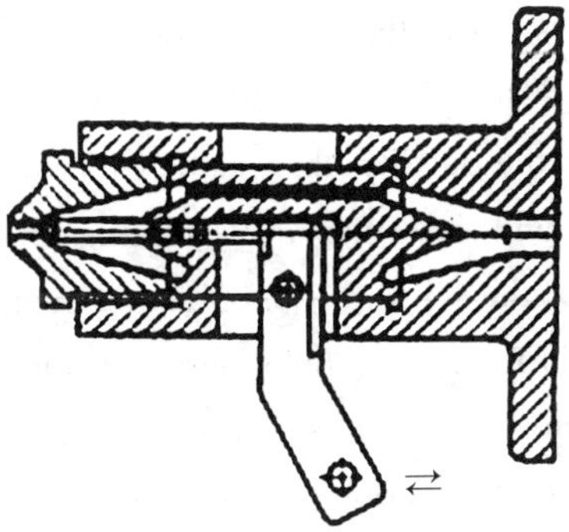

Fig. 3 - 9　Hydraulic control lever needle-type nozzle

液控杠杆针阀式喷嘴

The selection of the nozzle should be considered according to the performance of plastics and the characteristics of products. For plastics of high melt viscosity

聚氯乙烯

and poor thermal stability, such as polyvinyl chloride, it will be better to use large-diameter straight-through nozzle whose flow channel resistance is small and shearing function is smaller. For plastics of low melt viscosity such as polyamide, in order to prevent flow casting, it is advisable to use self-locking or lever needle-type nozzle with a heating device. Thin-walled products of complex shapes are suitable for adopting the nozzle of small aperture and longr ange. However, thick-walled products have the best selection of the nozzle of large aperture with large feeding function.

混色喷嘴

筛板

In addition to these kinds of nozzles which are mentioned before, there are nozzles for special-purposes. E. g. color mixing nozzle, is to improve the plunger-type injection machine to mix the pigment and powder uniformly. The nozzle contains a sieve to increase the shear mixing action to achieve the purpose of mixing uniformly. Molding thin-walled products can use the point-injection type nozzle, which has short sprue, and directly contacts with the cavity, and its pressure loss is small, which is suitable for a good fluidity such as polyethylene and polypropylene. The design of nozzle should be as simple and easy for loading and unloading as possible. The diameter of the nozzle hole should be based on the maximum injection amount of the injection molding machine, the nature of the plastics and the characteristics of the products. The nozzle head is generally hemispherical, and it requires the concave spherical to maintain a good contact with the mold sprue bushing. The diameter of the nozzle hole should be smaller than the diameter of the sprue 0. 5 to 1 mm to prevent leakage of material and avoid dead ends, and it is also convenient to pull the accumulated cold material along with the sprue warts out of the holes between the two injections.

6. Pressure Device and Driving Device

液压力

Pressure device is the facility that makes the plunger or the screw do necessary reciprocation and injection cycle during the injection by exerting pressure of the plunger or the screw. Its power source includes fluid pressure force and mechanical force, and the most adopted force is fluid pressure force, and most of them adopt self-contained hydraulic systems to supply pressure.

交流电机

Driving device is the device which makes the injection screw rotate so as to complete the process of the pre-plasticization of the plastics. Commonly used driving devices are single-speed AC motor and hydraulic motor. Adopting the motor to drive, it can guarantee the stability of the rotational speed. The advantages of using the hydraulic motor are as follows: ① Transmission characteristic is softer; the start inertia is small and has a protective effect on the overload of the screw; ② It is easy to achieve the change of the rotational speed of the screw from stepless to the higher speed; ③ It has small size of gearing device, light weight, simple structure and so on. The current trend is the application of the hydraulic motor.

3.2.2 Clamping System

锁模系统

Clamping system is the system which works as the lock of mold, controlling the opening and closing of the mold and the ejection of products. The melt is usually injected into the mold at high pressure of 40 to 150 MPa. To keep the mold closed tightly and not to spill the material, there must be a sufficient amount of clamping force. The value of the clamping force (F) depends on the product of injection pressure (P) and projected area (A) which is perpendicular to the pressing direction, i. e. $F \geqslant P_A$. In fact, the injection pressure has a great loss during the molding process. In order to meet the requirement of clamping, clamping force needs to only be greater than the product of mold cavity pressure P and the projected area (A, including shunt projected area), that is $F \geqslant P_A$. Mold cavity pressure is usually 40% to 70% of the injection pressure.

投影面积

齿轮

蜗杆

The clamping structure should ensure that the mold could open and close flexibly, accurately, promptly and safely. According to the requirements of the process, it should have a cushioning effect when opening and closing the mold. The running speed of the template should be quick at first and then slow when closing mold, while opening the mold, the speed should be from slow to fast and then slow to prevent the mold and parts from damaging, to avoid strong vibration of machine, to adjust to topping out fabricated parts smoothly, and to achieve the aim of safe operation and extension of service life of the machine and mold.

缓冲作用

使用寿命

The maximum stroke to open and close the template determines the maximum thickness of the fabricated parts that the injection machine can produce. The stroke of the template is adjustable to adapt to different dimensions of the mold within the maximum stroke.

The template should have sufficient strength to ensure that the mold will not be deformed under frequent percussion of pressure in the molding process, which may affect the stability of the product size.

Commonly there are three forms of mold opening and closing and clamping mechanism:

往复运动

1. Mechanical Type

移动模板

This device generally depends on the motor through gear or worm wheel, worm deceleration driven crank or the mechanism of moving the crank by lever to realize the opening, closing and clamping role (see Fig. 3 – 10). This form is simple in structure, easy to manufacture, convenient to use and repair, however, due to the frequent start of the drive motor, it leads to the great load of launching, the frequency of shock and vibration, and it is noisy and owns easy wear-out parts, a short trip of template and other factors, so it is only suitable for small injection machine.

液压—机械组合式/曲肘撑杆机构

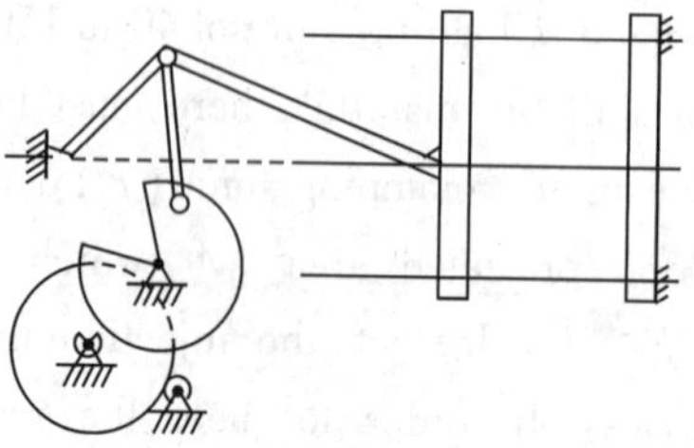

Fig. 3 – 10 Mechanical clamping device

2. Hydraulic Type

Hydraulic type is to use cylinder and piston by relying on hydraulic pressure to drive the reciprocating movements of piston to achieve the opening and closing of the mold and clamping, as shown in Fig. 3 – 11. Its advantages are as follows: ① Compared with other structures, the gap block between the mobile template and the fixed template is bigger; ② Mobile templates can be stopped anywhere in the range of travel stay, making it easy to install and adjust the mold, and easy to implement regulator and governor; ③ It works smoothly, reliably and is easy to implement emergent brake. But high power hydraulic system requires vast investment.

连杆式曲肘

运动特性

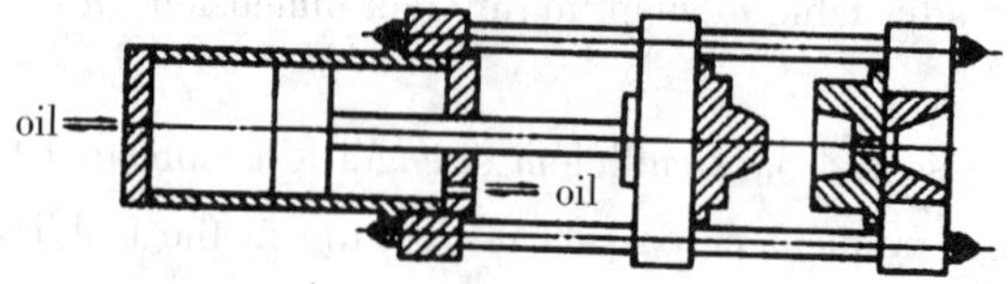

Fig. 3 – 11 Hydraulic clamping device

3. Hydraulic-mechanical Combined

顶出装置

This form is operated by the hydraulic linkage or the toggle strut mechanism to achieve the opening, closing and the locking of the mold, as shown in Fig. 3 – 12.

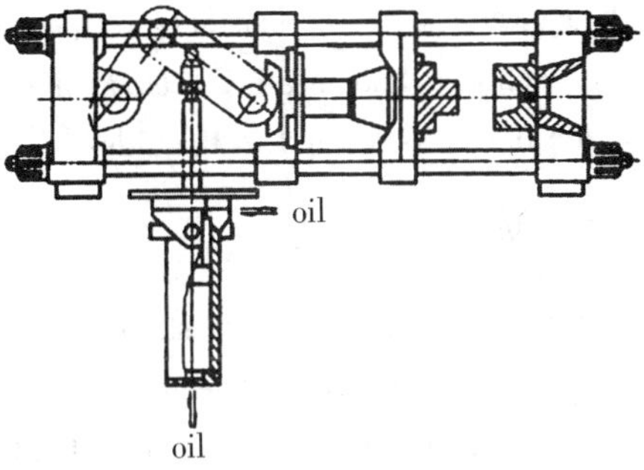

Fig. 3 – 12 Hydraulic-mechanical modular clamping device

The advantages of this structure are the following. ① Link-toggle owns self-increasing force. When it goes straightly, it has self-locking effect, even through removing hydraulic pressure, the locking force will not disappear. Therefore, the hydraulic system is set to handle only in the movement of rod or toggle, and the requiring load is not large so that it can save investment. ② The kinematic characteristics can meet the technical requirements, namely when the elbow lever promotes the template to close, the speed can be quick at first and then slow, however, when opening the template, the situation is opposite. ③ Clamping is more reliable. The disadvantage is that the structure is easy to wear out and relatively difficult to change transfer mode. It is still dominant in the application of small and medium-sized injection machine, and the key point is low cost.

The various clamping devices are equipped with ejectors , so that when the mold opens, products are ejected. Mechanical ejection and hydraulic ejection are two main types.

(1) Mechanical Ejection.

It uses the mobile top column which is located in the rack, to promote the mold release device, during the process of opening, which is located in the mold, to top out the products. This device is simple and more widely used. However, the ejection must be carried out in the opening time and the reset of the mold release device can be achieved only after the mold's closing. The characteristics of clamping determine the position of the top column and the release device, and they can be placed in the center or the sides of the template. Ejection distance can be adjusted according to different products.

(2) Hydraulic Ejection.

It depends on the hydraulic force of cylinder. Ejection force and ejection speed are both adjustable, but the top point is limited and its structure is more complex. For the large injection molding machines, the co-existence of these two kinds of machines is common and usually the acting cylinders are placed in the middle of the template and the mechanical ejection device is located in the sides of the template.

注塑模具

3.2.3 Injection Molds

Injection mold is the part which confers the shape and the size of the plastics in molding process. Although the structure of the mold might change according to varieties, performances, shapes, structures of the plastics, the injection molding machine type, etc., but the basic structure is the same. The mold is mainly composed of the pouring system, formed components and structural parts. Pouring system and formed components are in direct contact with the plastics, changing with different plastics and products, which are the most complex parts in modeling with the greatest change and the highest request of processing smoothness and accuracy. Typical injection mold structure is shown in Fig. 3 – 13.

浇注系统/成型零件/结构零件

Pouring system refers to the runner of plastics coming from the nozzle and then into the cavity, consisting of sprue, cold slug well, shunt, gate and so on. The formed components refer to various parts constituting the article shape, including the cavity, core, forming rod and so on.

分流道
冷料井/浇口
型芯/成型杆

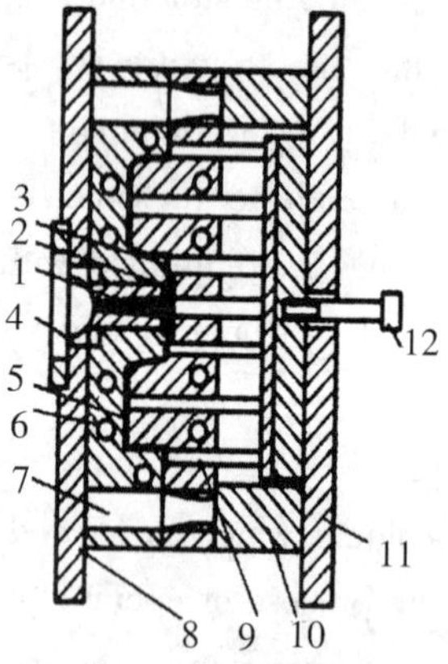

Fig. 3 – 13 A typical injection mold schematic diagram

(1—sprue; 2—shunt; 3—gate; 4—sprue bushing; 5—product; 6—cooling holes; 7—guide pin; 8—front clip template; 9—release lever; 10—bearing plate; 11—after the template folder; 12—screw)

1. Sprue

It is a passage in the mold which connects the nozzle of the injection machine with the diversion channel or cavity. The sprue is concave at the top so as to converge with nozzle. Sprue inlet diameter should be slightly larger than the nozzle diameter (0. 8 mm) to avoid overflow and interception because of incorrect convergence. The inlet diameter varies according to the product size, generally is 4 to 8 mm. The sprue diameter should be expanded inside, the angle of which is 3 to 5 degree to facilitate superfluous material in flow channel to demold. 凹形 赘物

2. Cold Slug Well

It is an empty hole located in the end of sprue to collect the cold materials generated in the nozzle end portion during two injections, thereby preventing the clogging of the gate or shunt. Once cold materials are mixed into the cavity, the product is easy to generate internal stress. The diameter of the cold slug well is 8 – 10 mm, and its depth is about 6 mm. In order to facilitate the stripping, the bottom is often supported by the putter, and the top of the putter should be designed into the twists and turns hooked or equipped with subsidence trench in order to smoothly pull the superfluous materials out of the sprue when stripping.

3. Runner

It is the channel which connects the main channel with various cavities in multi-slot die. In order to make melt fully fill the cavity at an equal speed, the arrangement of channel in the mold should be symmetrical and equally spaced. The shape and size of the cross section of the channel have an impact on the flow of the plastic melt and the release of the product , and it determines whether the mold manufacturing is easy or difficult. The resistance of the flow path of the circular cross-section is minimal considering equal quantity of current. However, the cylindrical flow passage must be opened in the mold halves and the channel is often trapezoidal or semi-U-shaped cross section for processing convenience. The flow channel surface must be polished to reduce the flow resistance to provide faster filling speed. 梯形/半 U 形

Runner dimension is determined by plastic variety, size and thickness of the product. For most thermoplastics, the shunt section width should not exceed 8 mm, and the maximum is up to 12 mm especially, and the minimum can be 2 mm. Under the premise of meeting the needs , it is necessary to minimize cross-sectional area, so as to avoid increasing diversion load and prolonging the cooling time.

4. Gate

It is the channel which connects the sprue (or shunt) with the cavity. The cross-sectional area of the channel can be equal to the sprue (or shunt), but usually it is narrow. Thus, it is the smallest part of the cross-sectional area of the runner system. The shape and dimensions of the gate have a great impact on the quality of the products. The role of the gate is to: ① control the flow velocity of the materials; ② prevent backflow due to the pre-cooling of this part of melt during the injection process; ③ make the melt which has already passed the runner to suffer from strong shear to increase temperatures, so as to reduce the apparent viscosity to improve liquidity; ④ facilitate the separation of products and pouring system. The design of gate shape, size and location depends on the nature of the plastic, and the size and structure of the products. Usually the gate cross-sectional shape is rectangular or circular and the cross-sectional area should be small and the length should be short, which is based not only on the above effects, but also on others, like it is easier for a small gate to become large while it may be difficult for a large gate to become small. The gate generally locates in the thickest part without affecting the appearance. The size design of gate should take the nature of the plastic melt into account.

流动性

5. Cavity

It is the space where plastic products are molded in the mold. The components used for forming the pocket collectively are called as molded parts. Each of the molded parts often has a special name; Molded part constituting the product shape is called die (also known as female mold), molded part which forms the shape of the interior of the article (such as holes, slots, etc.) is called core or punch (also known as the male mold). Firstly, the design of the molded parts should be based on the performance of plastics, the geometric shape of the products, dimensional tolerances and the requirements using to determine the overall structure of the cavity. Secondly, it selects the parting surface, the gate and a vent position and releases manner according to the confirmed structure. Finally, the design and combination between the various parts are decided by the size of product. Plastic melt goes into the cavity with a high pressure, so the molded parts should have reasonable selection of material and the check of the strength and stiffness. In order to ensure the smooth appearance of the surface which of the plastic products and release the mold easily, the roughness of the surface contacts with the plastics should be less than 0.63 μm with high corrosion resistance. Molded parts always use heat treatment to improve hardness and adopt corrosion resistant steels to produce.

凹模/阴模

凸模/阳模

6. Exhausting-slot　　排气槽

It is an outlet trough in the mold for discharging the original gas brought in by the melt. When injecting melt goes into the cavity, the original gas or brought-in-gas must be discharged at the end of the material flow to the outside through the exhausting-slot. Otherwise, products will have many problems, such as holes, poor welding and filling discontent. Even the accumulated air will burn products due to the high temperature of the compression of air. In normal situations, the exhausting-slot can be located in the end of melt flow of the cavity as well as the mold parting surface. The latter opens a shallow groove in the side of the dies with the depth of 0.03 to 0.2 mm and the width of 1.5 to 6 mm. During injection, exhausting-slot does not have a lot of melt leakages, because the melt will be cooled to solidify so that the channel is blocked. The opening position of exhausting-slot should not face to the operator, preventing the melt accidentally ejecting to hurt people. In addition, it is possible to exhaust gas through the operational gap between the putter and the putter hole, or the operational gap between the push board and core.　　推件板

7. Structure Part

It refers to the various parts constituting the mold structure, including the guide, stripping, core pulling, typing and various parting components, such as fixed mold fixing plate, moving mold fixing plate, bearing plates, guide posts, push plate, push pin and the return pin.　　支承板　回程杆

8. Heating or Cooling Device

It is the device which makes the material solidified in the mold. For thermoplastics, punch and die are equipped with a cooling medium channel to cool by circulating flow of the cooling medium. Cooling medium being introduced varies with the types of plastics and the structure of articles, such as cold water, hot water, hot oil and steam. The key point is high efficiency and uniform cooling because uneven cooling will directly affect the quality and size of the products. It should be based on the thermal properties (including crystallization) of the melt, the shape of the article and the mold structure, considering the arrangement of the cooling passage and the selection of the cooling medium.　　结晶

3.3 Process and Control Factors of Injection Molding

注射模塑工艺过程及控制因素

The injection molding process includes preparations before molding, the injection process and post-process, which are described as below.

成型前的准备

3.3.1 Preparations before Molding

In order to carry out injection smoothly and ensure product quality, equipments and plastics are prepared for the injection process.

1. Pretreatment for Raw Material before Molding

收缩率

According to the characteristics of a variety of plastics and the feeding situation, the general measures are to test the raw material appearance (color, particle size, uniformity, etc.) and technological properties (melt flow rate, flow resistance, thermal properties and shrinkage).

聚碳酸酯/聚砜/聚甲基丙烯酸甲酯/亲水基团

Some plastics, such as polycarbonate, polyamide, polysulfone, polymethyl-methacrylate, etc., their macromolecules contain a hydrophilic group, which is easy to absorb moisture, and leads to containing different degrees of moisture. If the water contents are higher than predetermined amount, it will cause the phenomena of silver, stripes, bubbles and other defects of product surface, and the worse result is causing degradation of raw material in injection process. It seriously affects the appearance, the quality of products and significantly reduces the performances of various indicators. Accordingly, this kind of plastics should be fully dried before molding. Non-hygroscopic plastics, such as polystyrene, polyethylene, polypropylene polyoxymethylene plastics, etc., generally may not be pre-dried if the storage and transportation and fine and packaging is tight.

聚苯乙烯
聚甲醛

Drying methods of a variety of plastics should be selected according to its performance and specific conditions. Small batch production of plastics, mostly are dried with hot air circulation oven or infrared heating oven; the plastics which is easily oxidized and discolored under the high temperature for a long time, such as polyamide, should be dried in the vacuum oven; plastic mass production should be adopted boiling-drying or air-drying, because the drying efficiency is high and continuous. The drying temperature, under atmospheric pressure, should be selected at 100 degree above. If the glass transition temperature of plastics is less than 100 degree, the drying temperature should be controlled to below the glass transition temperature. Generally, extending drying time will help improve the drying effect, but drying temperature for each plastic has a best drying time. If extending in drying time is mush, the drying result may be not noticeable. It is worthy of mentioning that the prevention from moisture of dry plastics should be paid attention to.

玻璃化温度

Some products with color need to add pigment or master batch and mix well before molding. If raw material is the powder, it is necessary to make it become granular.

2. Cleaning Sleeve

It is required to clean, demolish and replace the injection machine (main cylinder) before using some kinds of plastics or a certain injection machine for the first time, or changing the product in production process, replacing raw material and exchanging color or happening plastic decomposition phenomenon.

Cleaning sleeve of plunger-type injection machine is more difficult than that of screw injection machine because deposit quantity in the sleeve of plunger is large and not easy to rotate. So its sleeve must be removed or cooperated with a dedicated sleeve when cleaning.

Screw injection machine is usually directly reloading cleaning. To save time and material, it should take the right steps. Grasp the plastic thermal stability, compatibility and other technical information between the molding temperature range and a variety of plastics. E. g. exchange the plastic molding temperature lower than the sleeve temperature of the memory to remain plastics' high thermal stability while the thermal stability of pre-stored plastics is good. Firstly increase the temperature of sleeve and nozzle to the minimum processing temperature of the plastics, then add for refueling (for refueling material back) and continuous-to-air injection after all the holding materials have been cleaned by adjusting the temperature for normal production. When molding temperature of pre-changed plastics is far lower than the temperature of plastics in sleeve, the sleeve and nozzle temperature should be increased to the best current temperature of the plastic sleeve, then cut off the power supply, and clean with the pre-refueling in cooling condition refueling. For cooling cleaning, molding temperature of pre-refueling is high; melt viscosity is big; and retention within the sleeve material is heat-sensitive, such as PVC, POM or polychlorotrifluoroethylene, etc. To prevent the decomposition of plastic, good fluidity and high thermal stability of polystyrene or low-density polyethylene plastic for transition refueling should be used.

相容性

聚甲醛/聚三氟氯乙烯

3. Preheating Inserts

嵌件

For assemblage and requirements of usage, metal inserts should be often embed into the plastic parts. Prior to injection, the metal inserts should be put into a predetermined position within the mold, after molding makes it a whole member with plastics. Plastic product with inserts is easy to crack in the periphery of the inserts, which will cause the decrease of the intensity of the products due to the large differences of thermal properties and shrinking percentage between metal inserts and the plastics. So in addition to increase the wall thickness around the inserts in the design of parts to overcome this difficulty, preheating the metal inserts during forming is an effective measure. Differential temperature between the melt and inserts can be reduced after preheating, and the melt around the inserts cools slowly and contracts more uniformly during molding. Certain hot material feeding effect occurs, in order to prevent too much internal stress around the inserts.

刚性分子链
聚苯醚

Insert preheating depends on the nature of the processing plastics and the size of metal inserts. Plastic parts with a rigid molecular chain, such as polycarbonate, polysulfone, polyphenylether, etc., in molding are prone to appear stress cracking, so metal inserts ships should be preheated.

铬
铝

Small inserts which are easy to become plastic melt and heated in the mold do not need to be preheated. The preheating temperature is limited to 110 – 130 degrees, which will not damage the surface of the metal inserts that plated with zinc or chromium layer. The preheating temperature can be increased to about 150 degrees for aluminum alloy or copper insert without coating on the surface.

脱模剂

4. Selection of Release Agents

硬脂酸锌
液状石蜡
硅

The release agent is one auxiliary on the surface of the mold to make the plastic parts easily deposited from the mold. The release agents of general injection parts, mainly depend on the reasonable processing conditions and the right mold design. But in order to successfully release mold in production, release agent is commonly used. Commonly it has zinc stearate, for plastic, except for the polyamide plastic, liquid paraffin wax (also known as white oil) which is acting as a type of polyamide plastics, preferably by the effect of mold release agent, in addition to the lubricating, as well as the preventing parts voiding generated internally and silicone oil, which has good lubricating effect, but is expensive, more troublesome to use (formulated as a toluene solution, smear on the surface of the cavity, after heating and drying show an excellent effect), and has restrictions on using. Regardless of what kind of release agent, it should be in the right amount. Too little will not achieve the desired effect; excessive or uneven painting will affect the appearance and intensity of parts, for the transparent parts it is more obvious. Besides, too much amount will cause hair spot or turbid phenomenon.

毛斑/浑浊现象

注射过程

3.3.2 Injection Process

A complete injection process includes feeding, plasticizing, mold injection, stably cooling and stripping, but essentially it is divided into two processes: plasticizing, flowing and cooling.

1. Plasticizing

Plasticizing is the whole process when plastics is heated to flow in sleeve and has a good plasticity. It is said that the plasticization is the preparing process of mold injection. The general requirements for this process are: before entering the mold cavity, temperature should reach a predetermined molding one, and within the specified time a sufficient quantity and a uniform temperature of the molten plastics can be provided. Thermal decomposition should rarely occur in order to ensure that the production is carried out continuously. The above requirements are closely related with properties of the plastics, control of process conditions, and the plasticizing structure of injection machine, and also directly determine the quality and quantity of the work pieces. The issue of screw and sleeve plasticizer has been discussed in the previous chapter, here we only briefly discuss plasticizing of plunger-type injection molding machine.

(1) Heat Uniformity.

Due to the poor thermal conductivity of plastics, and its movements in the plunger-type injection machine only relying on the plunger's push, there is almost no mixing effect. It is detrimental to the heat transfer, so that the plastics temperature close to the sleeve wall is high, but is low in the cylinder center, forming uneven temperature distribution. In addition, when the melt flows in the tube, the material flow speed at the center of the cylinder is faster than that close to the cylinder wall inevitably. The different radial velocity distribution will further lead to the uneven temperature of each point of melt emitted by injection molding machine. Even the average temperature of the material of each shot is not even. Melt molding products have poor thermal uniformity, and their physical and mechanical properties are also poor.

The concept of heating efficiency (E_h) is introduced to analyze the thermal uniformity of the melt in the plunger-type injection machine. Set sleeve temperature as T_w, the temperature of the plastic's flowing into the sleeve as T_1. If the plastics stays long enough in sleeve, all the plastics temperatures will rise to near T_w, and T_w is the upper temperature limit, then the maximum distance of plastics' temperature rise is $T_w - T_1$, which will be directly proportional to the maximum heat obtained by plastics. But plastics' average temperature emitted by the nozzle T_2 is usually lower than T_w, so the average distance of actual 加热效率 平均距程

temperature rise of the plastics is $T_2 - T_1$, and the heat actually obtained is also proportional to this process. The ratio of two distances is the heating efficiency E_h.

$$E_h = \frac{T_2 - T_1}{T_w - T_1} \tag{3-1}$$

It must be pointed out that if the plastics in sleeve stays long enough, and also gets friction heat, T_2 will be greater than T_w, then E_h is greater than 1. But when plunger-type injection molding machine injects plastics with poor melt viscosity, this phenomenon is rare.

As mentioned above, the temperature of each point of the plastics emitted by the injection nozzle is uneven, its maximum temperature limit is T_w. Now assuming that the minimum temperature T_3 ($T_3 \geqslant T_w$ or T_f), T_3 must be higher than T_1 and less than T_2. Therefore, the actual plastic's temperature distribution ranges from T_3 to T_w. If the distribution of the plastics' temperature range is smaller in T_w fixed case, T_2 is higher, E_h is greater, as shown in Fig. 3 – 14. E_h not only indirectly says the size of the T_2, but also says the plastics' thermal uniformity. In production, plastics' injection temperature is neither below its softening point, nor higher than the decomposition humidity, so T_2 has a range in size. Practice has proved that when the E_h value is above 0. 8, the quality of the products is acceptable. Accordingly, when T_2 has been set, T_w is not difficult to determine.

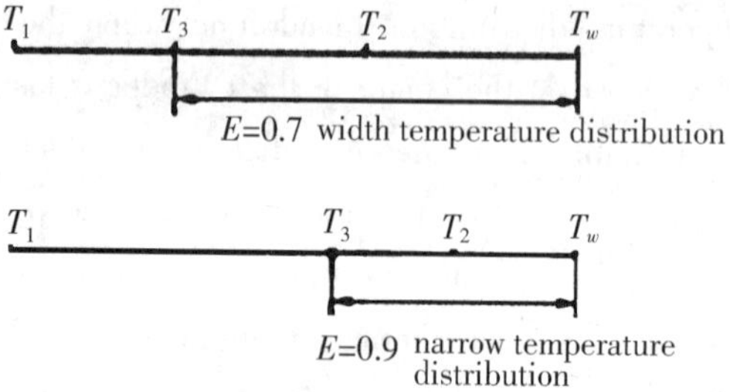

Fig. 3 – 14 The relationship of heating efficiency and temperature uniformity

(T_1—the temperature when the plastics into the barrel; T_2—the average temperature of the melt; T_3—the lowest temperature of melt; T_w—cylinder temperature)

Obviously, the value of E_h depends on the structure of the sleeve, residence time of plastics in the sleeve and the thermal conductivity properties of the plastics. Functions of this relationship is expressed as follows:

$$E_h = f\left[\frac{at}{(2a)^2}\right] \tag{3-2}$$

In the formula a is the thermal diffusion rate; t is plastics' residence time in the sleeve; a is the thickness of heating material. If shunt spindle is also used as heater, formula (3 - 2) becomes: 热扩散速率

$$E_h = f\left[\frac{at}{a^2}\right] \tag{3-3}$$

(2) Plasticizing Amount. 塑化量

Plasticizing amount refers to the quantity of plastics melted in the injection machine per unit time. The amount of plastics of plunger-type injection molding machine q_m is shown in formula (3 - 4):

$$q_m = K\frac{A^2}{V} \tag{3-4}$$

In the formula, A is the heated area of the plastics; V is the volume of heated plastics; K is a constant (with fixed injection machine, plastic, average temperature of the plastics and Ea for the premise). 受热面积

On analysis, if we want to increase the amount of plasticizing q_m, it is beneficial to increase A and reduce V. But A is limited by sleeve structure of plunger injection machine, so increasing A is bound to increase V. The method to resolve this contradiction is to use shunt spindle for dual-purpose as heater and changing the shape of shunt spindle. With the same plastics but different injection machine for injection, if the average temperature and the heating efficiency of emitted melt are fixed, the value of K can be used as the benchmark of the sleeve design.

Fig. 3 - 15 shows the temperature rise curve of the plastics in the sleeve from the feeding part to the nozzle. From the figure, in the plunger injection machine, we can see the plastic temperature rises faster near the cylinder wall, but slower in sleeve center; material temperature is not rising rapidly until it flows near shuttle through the shunt, and gradually decreases the temperature difference between the points and the plastics. But the final material temperature is still below the cylinder temperature. In screw injection machine, plastic heating rate is slower than the material near the sleeve wall of ram machine (this is due to plastic axial displacement under the action that the screw is faster than the plunger-type injection machine), but due to the screw mixing and shearing action, which not only can provide a lot of friction heat, but also can accelerate the transfer of the outer heating, in addition to the thin thickness of the plastic material layer, the material temperature rises soon. If there is strong shear, the material temperature may rise to sleeve temperature, or even go higher than the sleeve temperature before it reaches the nozzle.

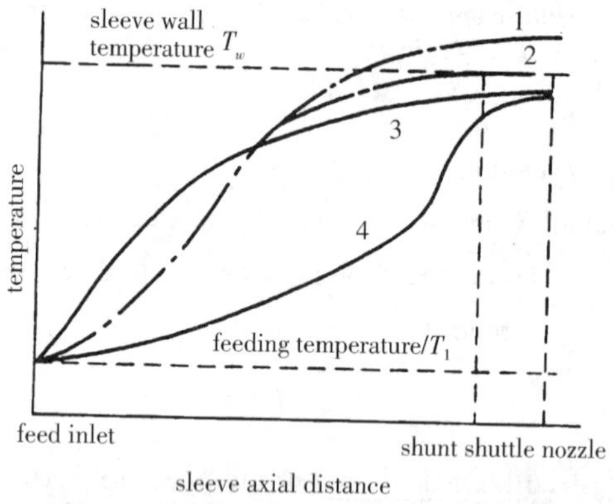

Fig. 3 - 15 Plastic heating curve in injection machine barrel

(1—mobile screw injection machine, shearing strong; 2—mobile screw injection machine, shearing more gentle; 3—plunger-type injection machine, near the barrel wall material; 4—plunger-type injection machine, the material of the center part)

2. Flowing and Cooling

This process begins from the plastic melt with uniform temperature and fluidity is injected into the mold driven by plunger or screw, filling the cavity, cooling and fixing under the controlled conditions until products demold from the mold cavity. It costs little elapsed time, but the melt changes a lot in the meantime, and the change has important implications for the quality of the products.

Melted material needs to overcome a series of flow resistance (including the melt, sleeve, nozzle, injection system, the cavity of the outer friction and melt internal friction) from the sleeve to mold cavity. At the same time, it also needs to compact the melt. Therefore, the injection pressure is high.

When the plastics is under pressure and heated in the plunger injection machine, firstly, it is pressed into the pillar-shaped solid from granular material by the pressure, and then it gradually becomes semi-solid or the melt during heating. So, there are three states or three sections for the plastics in sleeve. Flow resistances of the three sections in the injection are different. Resistance of columnar solids in the flow can be presented as pressure drop Δp_s.

半固体

$$\Delta p_s = (1 - e^{-\mu L/D}) p_M \qquad (3-5)$$

In the formula, e is the natural logarithm base; μ is the coefficient of the friction between the particulate solid and the wall; L is the length of the particulate solids within the tube; D is diameter for tubes; p_M is pressure to promote particulate solid or injection pressure. When injecting pressure, the loss of speed segment can be as high as 80% of the total pressure loss in the sleeve. The semi-solid and melt pressure loss can be calculated by the following formula (3 - 6):

$$\Delta p = \frac{2L}{R}\left[\frac{(m+3)q}{\pi k R^3}\right]^{\frac{1}{m}} \tag{3-6}$$

In the formula, R is the radius of the tube; q is the volume rate; k is the flow constant; q is the function of difference degree with Newtonian fluid.

It can be seen from the formula (3 - 5) and (3 - 6) that the pressure loss in three states decreases with the increasing diameter of cylinder. It is harmful to increase the diameter for the plasticizing amount, so there is a contradiction between the plastic flow and the heating process in the plunger injection machine.

When injecting, plastics in mobile screw injection machine encounters two types of resistance: the first one is the fluid flow resistance between the top of the screw and the nozzle. But it must be noted that because the plastic temperature of this section has reached the molding temperature, the role of sleeve outer heating is only insulating rather than heating, there is no contradiction between the plastic flow and heat. Sleeve's diameter can be large. The second one is that, the length of this region will not be too long, because the total quantity is only slightly larger than an injection volume, and average material temperature has reached optimum value, and the viscosity is lower too. To summarize, resistance in this section is smaller than the melt section of plunger injection machine melt. The second type of resistance is the resistance between plastics in the screw and inner sleeve. No matter whether the plastics is the solid, semi-solid or melt, its flow resistance can be obtained from formula (3 - 7).

$$F_f = \mu p A \tag{3-7}$$

In the formula, F_f is the frictional resistance; μ is the friction coefficient between the plastics and sleeve; p is the pressure of plastics; A is the contact area of plastics and tubular wall. 摩擦阻力

In screw injection machine, because the plastics melts faster, the solid zones will not be very long (i.e. A is not great). While the pressure (p) is not large, the resistance in the solid area is small. In the fluid and semi-solid region, plastics closed to the cylindrical wall has been melted, and the value of μ is significantly reduce, and also the value of A is not large, so the resistance is still small.

From the above analysis, the injection resistance of mobile screw injection molding machine is smaller than that of the plunger-type injection molding machine.

Regardless of what type the injection machine is, the flow of plastic melting into the mold cavity can be divided into four stages, including mold filling, compaction, back flow and cooling after gate freezing. In four successive stages, the temperature of the plastic melt will continue declining. The pressure change is shown in Fig. 3 - 16.

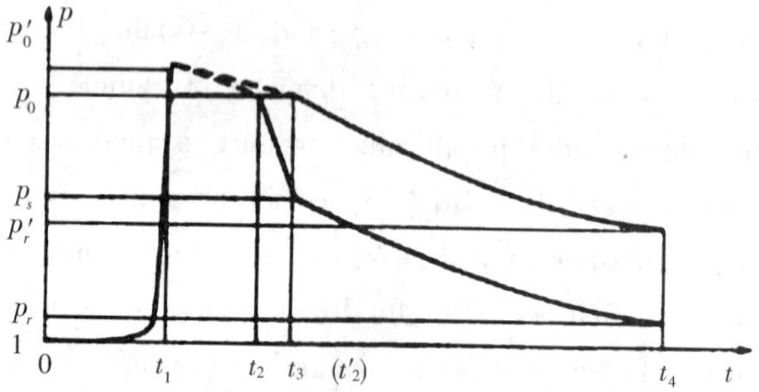

Fig. 3 - 16 Pressure changes in the plastic molding cycle

(p_0—molding maximum pressure; p_s—the pressure when gate is frozen; p_r—the residual pressure when sintered; t_1, t_2, t_3, t_4—each represents a certain time)

充模阶段

(1) Mold Filling Stage.

This stage begins from the plunger or screw moves, forwards until the cavity is filled with plastic melt (from zero to t_1). At the beginning of filling, there is only atmospheric pressure within the mold cavity. When the mold cavity is filled, pressure of material flow rises rapidly and reaches the maximum value p_0'. The mold filling time has something to do with molding pressure. If the mold filling time is long, when the plastics goes into the mold, at first it will be cooled more sufficiently, and the viscosity will increase, then the following plastics needs higher pressure to enter into the mold. Conversely, the required pressure is small. In the former case, as the plastics gets high shear stress, the degree of molecular orientation is relatively large. If this phenomenon remains until the material temperature lowers to the point below the softening point, directed molecular frozen in products will make products anisotropic. Products of this kind will crack when used in process which has great temperature change. The direction of cracks and the direction of molecular orientation are the same. Moreover, the thermal stability of the products is also poor, and this is because the softening point of the plastics reduces with the higher degree of molecular orientation. Filling mold in high-speed, melt will produce more friction heat and increase the temperature of material when getting through the nozzle, sprue, runner and gate. When the pressure reaches the maximum value at t_1, the temperature of the plastic melt will be able to maintain a higher value, then the degree of molecular orientation can be reduced; the article weld strength can also be improved. However, when the mold cavity has the core or insert of split stream, if the mold is filled too quickly, in the rear of the core or insert, a lower strength of the weld marks will be produced, resulting in the deterioration of the product intensity.

(2) Compaction Stage. 压实阶段

It refers to the period of time since the melt fills in mold cavity until the plunger or screw is withdrawn (from t_1 to t_2). During this time, the plastic melt will occur contraction due to cooling, but because the plastics is still in the regulator pressure of plunger or screw, melt within the sleeve will continue to flow into the mold to replenish the gap leaving by shrinking. If the plunger or screw stops in place unmoved, the pressure curve attenuates slightly and drops from p'_0 to p_0; if plunger or screw keeps the pressure constant, that is to say, with the entering of molten material into the mold, plunger or screw makes a slight moving forward at the same time, then the cavity pressure in this section remains unchanged, and the pressure curve and the time axis are parallel. Compaction stage has impact on improving the density, reducing shrinkage and overcoming the surface defects of the products. In addition, since the plastics is still in current, and the temperature is still declining, and directional elements are easy to freeze, this phase is the main stage of orientation forming of the macromolecular. The longer the delay of this stage is, the greater the degree of molecular orientation will be.

(3) Back Flow Stage. 倒流阶段

This stage is starting from the plunger or screw retreat (from t_2 to t_3). In this moment the pressure in mold cavity is higher than the flow path, so back flow occurs easily, then the pressure in mold cavity decreases rapidly from p_0 to p_s. Back flow will not stop until the melt at the gate freezes. If the melt at the gate has been frozen when plunger or screw retreats, or a non-return valve is equipped in the nozzle, then the reflux stage does not exist, and will not appear pressure drop curve at the segment of t_2 to t_3. So whether the reflux is much or little is determined by the time of press forming phase. However, no matter the frozen of melt at the gate is before or after the plunger or screw retreats, the pressure and temperature when freezing are always important factors to determine the average shrinkage of products, and these factors are impacted by the time of compaction stage.

Back flow stage has the flow of plastics, and therefore it will increase the molecular orientation, however, this orientation is relatively rare, and the spread area is not large. On the contrary, since the plastic temperature in this stage is high, some directed molecules may also discharge because of Brownian motion.

布朗运动

(4) The Cooling Stage After Freezing.

This stage starts from the time when melt at the gate freezes completely to the time when plastic products have been ejected from mold cavity ($t_3 - t_4$). The pressure in the mold cavity declines rapidly, from p_s or p_0 to p_r. Plastic in mold at this stage is to continue cooling, so that the products can have sufficient rigidity and not cause distortions in the stripping. In this stage, although there is no plastic flows into the gate, in the mold there may have a small amount of current, therefore, a small amount of molecular orientation still can be produced. As the temperature, pressure and volume of plastics in mold change at this stage, the pressure within the mold is not necessarily equal to the external pressure until the products release. The different value between pressure within the mold and the pressure outside is called the residual pressure. The residual pressure is closely related to the time length of compaction stage. When the residual pressure value is positive, stripping is more difficult and the article is easy to be scratched or broken; when residual pressure value is negative, the surface of the article is easy to dent or has internal vacuum bubble. Therefore, only when the residual pressure value is close to zero, the stripping is smooth, and we can get satisfactory with the products.

变形

残余压力

真空泡

It should be noted that the plastics is cooled from getting into the mold until the mold releases. If the cooling is agent and the temperature of all contact parts between mold and plastic is different, then the uneven cooling will lead to uneven shrinkage, and then the products will produce internal stress. Even uniform cooling, the rate when the plastics getting through the glass transition temperature in the cooling process may also be faster than the rate of molecular conformational transition.

分子构象转变速率/制件的后处理

3.3.3 Post-treatment of Products

After stripping or machining, injection products often need to be carried out an appropriate post-treatment to improve and enhance the performance and dimensional stability of the parts. Post-treatment of products refers to the annealing and humidity treatment.

退火

调试处理

1. Annealing Treatment

The uneven plasticizing of the plastics in sleeve or different cooling rates in the mold cavity often can produce uneven crystallization, orientation and contraction, and this will lead to internal stress. It will be more prominent in the production of products that are thick-walled or with metal inserts. The work piece with internal stress usually leads to the deterioration of mechanical properties, surface crazing, the decline of optical properties, or even deformation and cracking in the process of storing and using. Method to solve these problems in production is to process parts by annealing treatment.

力学性能
光学性能

The annealing process allows the products to stand for a period of time in the constant temperature of the heated liquid medium or a hot air circulating oven (such as hot water, hot mineral oils, glycerol, ethylene glycol, liquid paraffin, etc.). Processing time depends on the varieties of plastics, the temperature of the heating medium, the shape of the article and molding conditions. The plastics which has more rigid molecular chains, greater wall thickness with metal inserts, wide temperature range, high dimension precision, high internal stress and less self-elimination requires to be annealed. However, for the plastic parts of paraformaldehyde and chlorinated polyether, there is an inner stress, but because the molecular chain to lee itself is flexible and glass transition temperature is lower, the internal stress can disappear by itself. If the requirements of the article aren't strict, annealing can be canceled. The annealing temperature should be controlled above the temperatures of products 10 – 20 degrees, or lower than the plastic heat distortion temperature 10 – 20 degrees. Too high temperature will make the product curve or deformation; too low temperature will not achieve to the purpose of eliminating internal stress. The annealing time depends on the thickness of the products, and it is appropriate when achieving the purpose of eliminating internal stress. When the annealing time arrives, the products should be cooled slowly to room temperature. Cooling the products too fast, it is possible to re-ignite the internal stress and all effort will come to naught. The substance of annealing is to: ① enable the molecular chain to be forced to freeze to relax, and the solidified macromolecular chain segments turn to a random location, thus eliminating the inner stress of this portion; ② raise the degree of crystallization, stabilize crystalline structure, thereby enhancing the elastic modulus products and hardness of crystallization plastics, and decrease the elongation at break.

介质
矿物油/甘油
乙二醇
氯化聚醚
断裂伸长率

2. Humidity Treatment

The polyamide plastic products often occur oxidation or discoloration when getting contact with air at high temperature. In addition, they are easy to absorb moisture and expand when used or stored in the air, then a long period of time is needed to get a stable size. Therefore, if we put article newly demolded in hot water, we can not only process annealing to prevent oxidation by isolating from the air, but also achieve absorbent balance quickly, so we call it as humidity conditioning treatment. The right amount of moisture also has plasticizing effect on polyamide, thereby improving the flexibility and toughness of parts, then impact strength and tensile strength are improved. Humidity conditioning treatment time varies with varieties of polyamide plastic, shapes of work piece, thickness, size and crystallinity.

注射模塑工艺条件的分析讨论

3.4 Analysis and Discussion of the Injection Molding Process Conditions

Production of high quality injection products involves many factors. In general, after proposing a new product performance and other relevant requirements, the selection of the most suitable raw materials, production methods, production equipments and mold structures should be after the principle of feasible technology and reasonable economic.

Under these conditions determined, selection and control of process conditions is a major consideration. Injection molding is a composite element method. Clearly, the overall molding compound can reduce the assembly process (adhesive or weldability) and the number of molds, and it is economically reasonable. However, the overall design of complex often encounters different wall thickness requirements or the side notch groove structures. These structures cannot be achieved by usual injection molding, while the use of gas injection can solve this problem. Pumping gas to the thick wall of products, avoiding hole or deformation, the side notch groove may be solved by changing the product design.

加强筋

Due to the large parts and structural components lack of sufficient strength and rigidity, combination strengthens during the post-processing, reinforcing ribs, or structural foam molding to thicken the whole article. Thus, the mold structure becomes complicated or increases the quality of the products; the molding cycle becomes long. From an economic point of view this is a disadvantage. Adopting GAIM and setting a hollow rib structure, you can solve these problems.

气辅注塑/中空筋结构

3.4.1 Application in the Forming of Sandwich Structure Material

This year new method which combines the advantage of GAIM and the forming of sandwich structure material opens. This method is to produce composite products of hollow thick-wall and two-component. This method first injects two kinds of materials into the mold partly, injecting gas at the moment of completing filling to complete the process of gas injection. Two-component composite products of thick-walled hollow obtained in this way are light, often with a soft outer layer, the inner hard materials for the handle, the handle, meeting the need for local enhancement of the products.

3.4.2 Application of GAIM in Structural Foam Molding

It maintains the advantages of gas-assisted injection molding, but the hollow parts of products are replaced with a foam core parts. The specific method is the material filling, gas injection, the control of gas pressure, to reduce pressure at a certain rate, so that the inner melt foams and expands to the center, to form two-component structural foam used for light weight and rigid products. 发泡芯材

Chapter 4
Calendering Mold

压延成型

Calendering mold is a method which makes plastic material become specified dimensions of the continuous thin film and calendered sheet by rolling and extending through the gap between a series of opposite revolved parallel rollers. Thermoplastics can be shaped by calendering mold; the products can be thin film, sheet, artificial leather and various kinds of coating products. Calendering mold is a main method to produce films and sheets. Calendered film products are mainly used for agricultural filming, industrial packaging, interior decoration and all kinds of daily necessities. Calendered sheet products commonly are used as floor, soft and hard disc substrate, conveyor belt, thermoforming, laminated sheet, etc. Calendered film thickness is in the range of 0.05 – 0.3 mm; sheet thickness is in the range of 0.3 – 1.0 mm. If thickness of product is less than 0.05 nm or more than 1.0 mm, generally, we do not use the calendering mold, instead, we use the methods such as blow molding or extrusion.

辊筒

人造革

软硬唱片基材

吹塑

Sheet forming is the linkage assembly forms with rolling process as the center. It has high degree of automation, and belongs to the continuous operation progress. Calendering mold has the advantages of rapid forming, high production efficiency, high precision, good quality, etc. An ordinary four roll calender processing capacity is 5,000 – 10,000 t of a year; line speed during the production of films is 60 – 100 m/min, and even up to 300 m/min. Rolling products have uniform thickness, and thickness tolerance can be controlled within 10%.

压延机

The main disadvantages of calendering mold include the large equipment, too many auxiliary equipment, high precision, high investment, complex repair, product width restricted by the roller length of compression molding machine, etc. So in the production of continuous extrusion, calendering mold develops slower than forming sheet.

4.1 Calendering Mold Equipment

压延成型设备

Calendering mold device comprises calender and auxiliary equipment. Different rolling products need to use different auxiliary equipments.

辅助设备

4.1.1 Calender

1. Classification

According to the number of roller, calendar is divided into double roller, three, four, five, and even six roll calender that is open rubber mixing machine or roller press, usually only for plastication, mixing and tabletting, not used in calendering mold. At present, rolling machine with three rollers or four rollers is most widely used. Rubber generally uses three rolling mill; plastic calender usually uses four roll calender. Five-roll and six-roll calender are used less. The more rollers the calender has, the more rolling times materials receive. Thickness of rolling products can be thinner and more uniform, surface more smooth, and the production efficiency can be higher, e. g. rolling speed of four-roll calendar is 2 – 8 times of the speed of three-roll calender, but the bigger the roller number is, the more giant and more complex the equipment will be. At present, four-roll calender is gradually replacing the three-roll calender, but five-roll and six-roll calenders are not commonly used.

Roller arrangement has triangular, linear, inverse L shape, Z shape, L shape, oblique Z shape. Several as shown in Fig. 4 – 1, are the most common ones. Different arrangements have difference in accuracy, operation, convenience of main-tenance, and the possibility of accessory device setting, therefore calendars with different arrangement modes should be suitable for different types of products. At present, the most common application is oblique Z shape and inverse L shape. Calender roller of the oblique Z shape arrangement can prevent the material from thermal decomposition. Each roller is independent and stable transmission, and roller spacing adjustment is easy. Feeding, production control, observation of examination and maintenance is convenient, but the material roll degree is low and the smooth degree of product surface is low. The type of this kind of rolling machine is particularly suitable for fabric joint, such as PVC artificial leather, etc. When the inverse L shape calendar machines are in production roller stress is small; the roller deflection is small; products with the

三角形

挠度

uniform thickness can be rolled out as long as to compensate the deflection of the fourth roll; the package degree of material is high, and surface finish degree of product is high, so it is particularly suitable for the production of thin and transparent film. It is the main equipment for the production of PVC films at present.

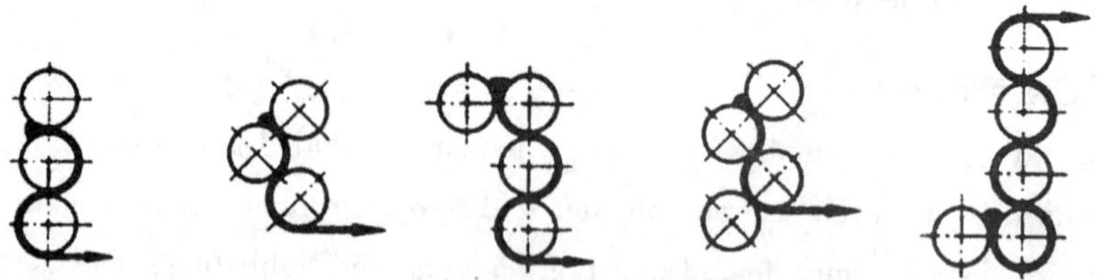

Fig. 4 - 1 Several calendar roller arrangement

2. Rolling Machine Structure

Although there are many types of calender, the basic structure is the same. Roller is the most important part of rolling machine, in addition, there are the base, frame, roller spacing adjustment device, driving device, heating and cooling system, lubrication device, shaft cross device, prestressed device, etc. Calender structure is shown in Fig. 4 - 2.

辊距调节装置
润滑装置

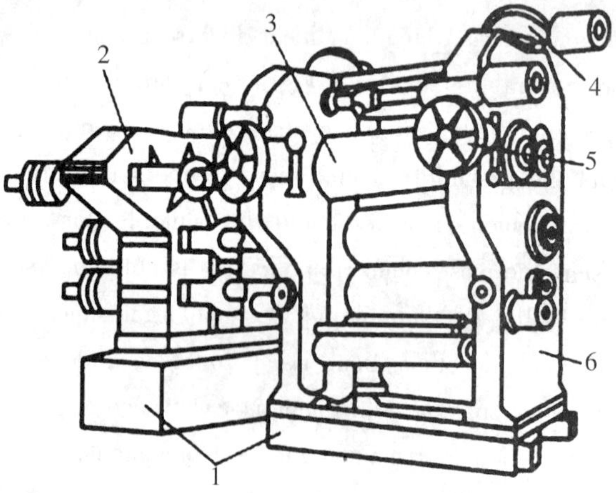

Fig. 4 - 2 The structure of calendar

(1—base; 2—transmission device; 3—roller; 4—roller distance adjustment device; 5—axis crosses regulating device; 6—frame)

(1) Roller.

Cylindrical roller includes cylindrical and drum-type. The hardness of roller surface is high(for example Durometer-hardness is 40 – 65) ,and it will be higher when they are used for rolling hard and high viscosity polymer products. Roller is mainly made of chilled cast iron, cast steel or alloy.

肖氏硬度

Inside the roller, steam, hot water and cold water get through and control the surface temperature. The structure includes hollow and hole type, as shown in Fig. 4 – 3. Cylinder wall of hollow roller is thicker; the heating area is small; the heat transfer is slow; the temperature difference of roller surface is bigger; central temperature is often higher than that of the end, resulting in uneven thickness of rolling products, so the adoption of hollow roller is less. Drilling roller is near the surface along the distribution around the drill with a through hole diameter of about 30 mm; the hole and the center hole is the same. The surface of heat carrier channel is close; heat transfer area is large; distribution of heat transfer is uniform, so the roller surface temperature is uniform, which can improve the precision extension of product and thickness uniformity of calendering products effectively. At present, the large high-speed rolling machine often uses drilled roller.

钻孔式压延辊筒

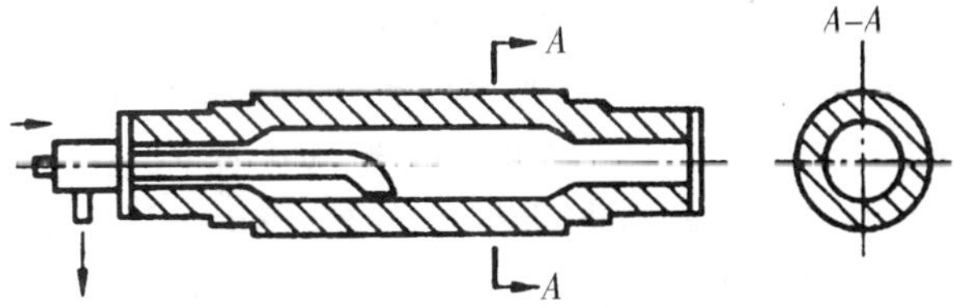

(a) hollow-type roller

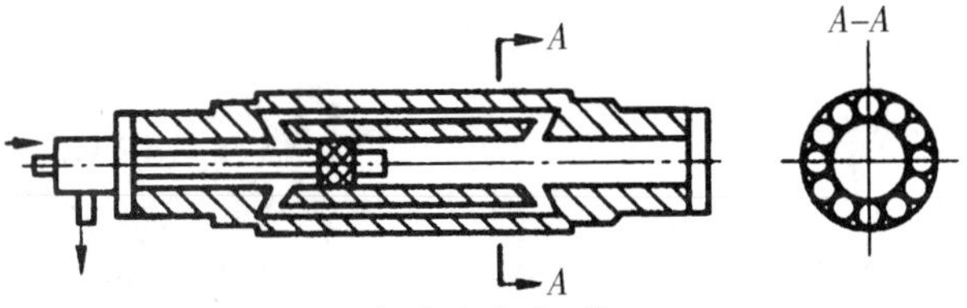

(b) drilled roller

Fig. 4 – 3 Roller structure

Calender roller commonly uses length and diameter to express specifications. Roller diameter determines the rolling velocity; roller length determines the width of products. Roller diameter cannot increase limitlessly, or rolling separating force is too large; at the same time in order to ensure that the roller has sufficient rigidity, the ratio of length to diameter of roller has certain restrictions. Usually, ratio of roller length to diameter should be smaller when used as soft products calendaring(2 – 2.2). Calender diameter of industrial production is usually 200 – 900 mm; working face's length is 500 – 2,400 mm. At present, the development trend of calender roller is to increase roller's width without increasing the diameter, even with unequal diameter roller.

(2) Other Parts.

The base is used for fixing the frame, usually made of cast iron. The frame locates in the two sides of the base for supporting roller spacing adjustment device, driving device, shaft cross device, prestressed device, etc.

The role of roller distance adjustment device is to regulate the roll of roller spacing according to the thickness of product requirements. Roller distance adjustment device usually sets coarse and fine. Roller distance adjustment device can stop automatically when the roll gap reaches a critical value (usually 0.1 mm), avoiding the damage to the surface of the cylinder.

临界值

Drive control device's role is to provide power for the calendering drum, using DC motor respectively to drive each roller to adjust the rotor speed and the roller speed ratio easily.

The lubricating device is composed of pump, pipe, heater, cooler, filter, oil tank, etc. It's role is to lubricate roller, reduce wear, and ensure the roller smooth and free operation. Shaft cross device and prestressed device's role is to overcome and compensate the bending and elastic deformation roller received from the calendering reaction, and reduce or overcome the problem of uneven transverse thickness of thin film.

磨损

弹性形变

4.1.2 The Auxiliary Equipment

辅助设备

Auxiliary equipment is divided into rolling before feeding equipment and rolling after the auxiliary equipment. Rolling before feeding equipment usually includes metering device, high-speed mixer (or kneading machine), mixer, double roller machine (or planetary screw extruder, both are kneading machine), metal detection device and material conveyer belt. Rolling after auxiliary equipment usually includes peeling-off roller, ginning device, cooling device, the thickness measuring device, cutting device, etc. Here, we do not talk too much about the calendaring before feeding equipment, only introduce rolling after auxiliary equipment briefly.

密炼机

行星螺杆挤出机

1. Peeling-off Roller

引离辊

The role of peeling-off roller is to peel off the shaped film without buckling and uniformity from calender roller, at the same time stretch the products for a certain degree, then pass to the embossing device. Sometimes a peeling-off roller with small diameter is disposed between the calendering rollers to reduce the effect of drawing the film. Length of the peeling-off roller and the length of calender rollers are almost the same in the middle, through steam heating in order to prevent the film away from the cold when drafting and plasticizers diverted from the roll being in precipitation. In order to reduce the rolls shrink when product leaves the peeling-off roller of the calender, distance between calender rollers should be as close as possible. Generally distance from the last calender roller is 70 – 150 mm.

2. Ginning Device

轧花装置

The role of ginning apparatus is not only to limit the pattern on the rolling surface of the article, but also to use so-called plain rollers which are highly polishedand chrome-faced (no pattern) to press polish to improve the brightness of the product surface. Ginning device consists of a crimped steel roller and a rubber roller. Cavities are cooled by water. Rubber roller is an active roller; crimped steel roller is a driven roller.

橡胶

3. Cooling Device

After stripping roll and ginning device, product temperature is high and needs special cooling by cooling device for setting. Generally, we take the method of gradual cooling for cooling products. The cooling device is mainly composed of 4 – 8 cooling rollers. Cooling roller is not suitable for chrome, but it is best to use aluminum grinding roller, to avoid the film's adhesion. Cooling roll generally uses two segments to control itself, so the cooling roll should be controlled properly to avoid cold drawing of product.

剥离辊

4. Thickness Measuring Device

测厚装置

Detection device for measuring the thickness applies a mechanical contact type, radioisotope method, electricity induction method, etc. Application at present most is the radioactive isotope thickness gauge, the measurable thickness of which is 0.10 – 2.10 mm, and the accuracy of measurement error is ± 0.005 mm.

放射性同位素/电感应

In addition, it also includes a cutting device, coiling device, inspection device, the use of artificial leather rolling of the drying roller, preheat roller, laminating device, etc.

贴合装置

压延成型原理

4.2 Principle of Calendering Mold

Calendering mold process is a process of material to produce viscous flow deformation in the extrusion effect of calender roller, so the material force and deformation rule in calendering mold (material force, flow velocity distribution, the stress-strain and rolling effect) are the main contents of the principle of calendering mold.

4.2.1 Pressure Distribution of Material in the Vise District

Fig. 4 - 4 shows the extrusion of material into two opposite rotating rollers. The material is carried into roll gap by the friction of roller and squeezed, and deformation is generated. Materials from entering into the roller gap (materials begin to be squeezed from roll in the starting vise point) to leaving the roller gap (materials are no longer subject to squeezing from rolls in ending vise point) are being extruded by roller, and this area is called the center clamping point c; maximum pressure area of pliers residential point b is maximum pressure clamping point.

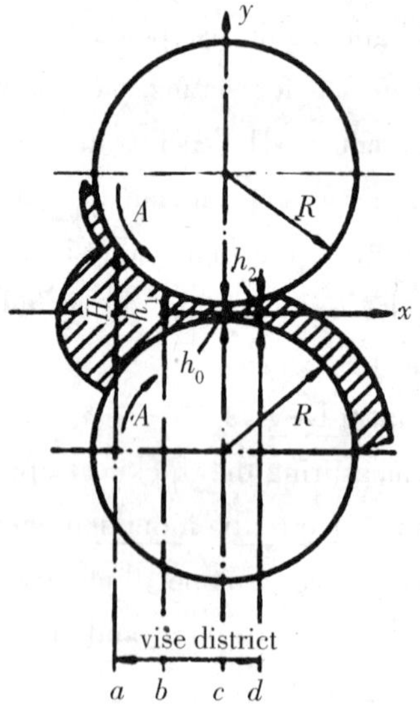

Fig. 4 - 4 The case when the material is squeezed between two rotating rollers

(a—starting vise point; b—maximum pressure vise point; c—center vise point; d—ending vise point)

In the direction of y axis, roller pressure of the material is the same everywhere. However, in the x axis direction of feeding in and ejecting out material, different position has different pressure. According to the fluid continuity equation and momentum equation, pressure equation is derived for the x axis direction as follows: 动量方程

$$\frac{dp}{d\lambda}=\frac{\eta v}{H_0}\sqrt{\frac{18R}{H_0}}\left[\frac{\lambda^2-\lambda_1^2}{(1+\lambda^2)^3}\right] \tag{4-1}$$

In the formula, p is the pressure of the material; η is the viscosity of material; v is the line speed of roller surface; H_0 equals to half of h_o (the gap between two rollers), R is the roller radius; λ is a function of the abscissa x; $\lambda = x/\sqrt{2RH_0}$ (zero dimensional variable); λ_1 is for embedded terminal point $x = x_1$ lambda values; $\lambda_1 = \sqrt{Q/2vH_0 - 1}$ (λ as zero dimensional variable, Q as the volume rate of flow unit roller width). 无因次变量

The formula (4 – 1) after integrating is combined with the boundary conditions (points a and $d_p = 0$) available:

$$p=\frac{\eta v}{H_0}\sqrt{\frac{9R}{32H_0}}\left[g(\lambda,\lambda_1)+5\lambda_1^3\right] \tag{4-2}$$

In the formula, $g(\lambda, \lambda_1)$ as complex functions of λ and λ_1 has two roots, point a and point d, at which the pressure value is zero.

The maximum pressure around the point of λ is $-\lambda_1$. By formula (4 – 2) we can further derive the maximum pressure to b stress:

$$p_{\max}=\frac{5\lambda_1^3\eta v}{H_0}\sqrt{\frac{9R}{8H_0}} \tag{4-3}$$

Pressure in x axis direction is varied since material enters into the roller gap to leave the roller gap. The theoretical calculation of the pressure distribution curve shown in Fig. 4 – 5 is in good agreement with the measured pressure curve, especially the pressure and the value of maximum pressure vise point are fairly consistent. 分布曲线

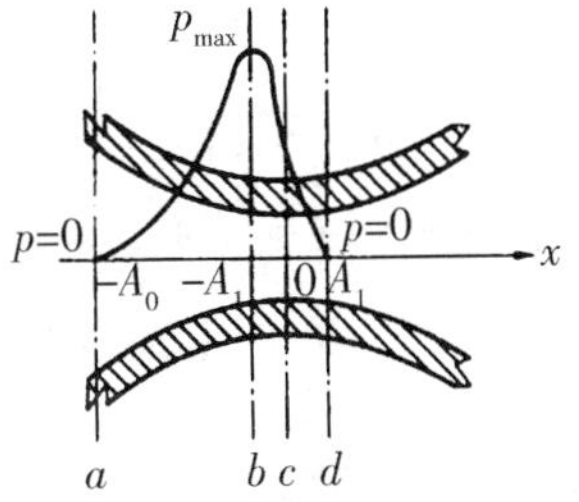

Fig. 4 – 5 Material in the roller gap; pressure distribution in x axis direction

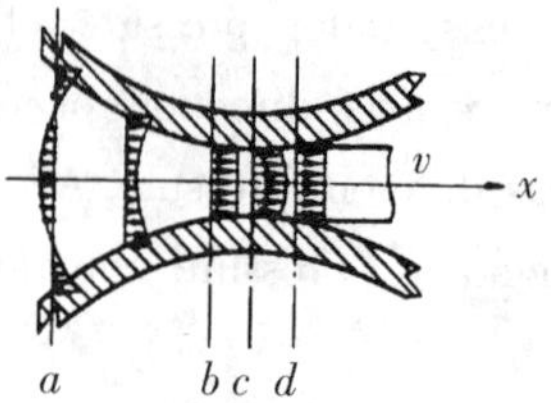

Fig. 4 – 6 Velocity distribution of material clamping zone of the roller

4.2.2 The Velocity Distribution of Material in the Vise Zone

In clamping area, material in two calender rollers gap flows through the dual role of roller friction and pressure, so the materials flowing along the x axis and y axis both have changes, that means velocity distribution exists both in the x and y axis.

Material flow velocity(v_x) at any residential location(x) of clamping area, and velocity distribution of(v_x) along the y direction can be given by:

$$\frac{v_x}{v}=\frac{2+3\lambda_1^2\left[1-\left(\frac{y}{h}\right)^2\right]-\lambda^2\left[1-3\left(\frac{y}{h}\right)^2\right]}{2(1+\lambda^2)} \qquad (4-4)$$

In the formula, v is for the speed of roller line, h for the distance between roller surface and x axis of clamping position area x, y for y coordinate values of clamping area of x's one point, λ and λ_1 for dimensionless variables.

The value of x is determined, then in formula(4 – 4), λ, λ_1 and h can be determined, because when the speed and roll diameter are certain, v is certain too. It can get velocity distribution curve along the y axis direction at any x location, as shown in Fig. 4 – 6. Seen from the graph, the material speed in point b and d is equal to the line speed of roller surface, and velocity distribution curve lines along the y axis direction are all straight lines. Pressure gradient value between point b and d is negative, and the velocity distribution is convex.

4.2.3 Shear Deformation Analysis of Material in the Vise

The formula(4 – 4) and partial derivatives with respect to y and $h=H_0(1+\lambda^2)$ formula of shear rate of material flow can be obtained in the vise zone, i. e.

$$\gamma=\frac{3v\lambda^2\lambda_1^2 y}{H_0^2(1+\lambda^2)^3} \qquad (4-5)$$

For the Newton fluid material, $\tau=\eta\gamma$, so the shear stress formula of material flow of clamping area is:

$$\tau = \frac{3v\eta(\lambda^2 - \lambda_1^2)y}{H_0^2(1+\lambda^2)^3} \tag{4-6}$$

According to formula(4 – 5)and(4 – 6), distribution of τ and γ in clamping area can be obtained, as shown in Fig. 4 – 7(a). Seen in the y axis direction, τ and γ are in linear distribution, in the clamping area of the neutral plane(i. e. $h_0/2$); τ and γ are zero; The value of τ and γ in the roller surface are the largest. At point b pressure is maximum in the clamping area(i. e $\lambda = -\lambda_1$), and at the end vise point d(i. e $\lambda = \lambda_1$), no matter what the value of y is, τ and γ are identically zero. τ and γ are both changing in the x axis direction.

Corresponding to the shear rate, shear deformation increases gradually after entering into the clamping zone, and becomes maximum at the maximum pressure point b. Since then deformation returns to a gradual recover, reaches the end point that is left in vise clamping area, and still retains some within the material, as shown in Fig. 4 – 7(b). So for viscoelastic material, the thickness of rolled products is always slightly larger than the last calender roll gap.

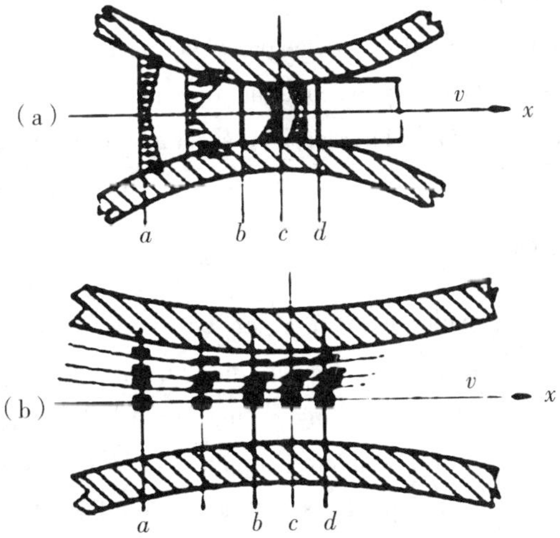

Fig. 4 – 7 Melt material between the roller shear stress and shear rate distribution and shear deformation of material schematic

4.2.4 Rolling Effect

压延效应

In the longitudinal and transverse direction physical and mechanical properties of rolling products are sometimes different, i. e. tensile strength is bigger along the rolling thin film or sheet longitudinal direction(rolling direction). Small elongation occurs, and shrinkage rate is high. If the tensile strength along the transverse direction(perpendicular to the rolling direction) is small, elongation is big, and shrinkage rate is slow. The different performances in vertical and horizontal direction of rolling products are called rolling effect. This phenomenon is mainly due to the arrangement of polymer chain or acicular, direction of flake filler particle rolling.

For rolling products for isotropic, calendering effect should be as far as possible to eliminate or be controlled to the suitable degree. If you want to introduce this orientation effect in the rolling products, such as films of high strength in single direction in the production process, measures can be taken to promote the calendering effect.

压延成型工艺

4.3 Calendering Mold Process

A variety of thermoplastics can be rolled by rolling process in principle. At present, polyvinyl chloride has the largest amount of rolling plastic varieties in the industry; what follows is polyethylene. Polypropylene modifies polystyrene and ABS. The main products produced by calendering process are film, sheet and artificial leather. Production of soft PVC film, rigid PVC sheet and artificial leather of foam PVC are the largest.

丙烯腈—丁二烯—苯乙烯

4.3.1 Calendering Mold Process

Entire calendering molding process can be divided into two stages, namely feeding stage (including material preparation, mixing, plasticizing and feeding processes) and rolling stages (including rolling, tracting, engraving, cool setting, conveying, winding or cutting and other processes). Fig. 4 – 8 is a calendering mold process of typical rolled products, including soft PVC film, rigid sheet and artificial leather.

刻花

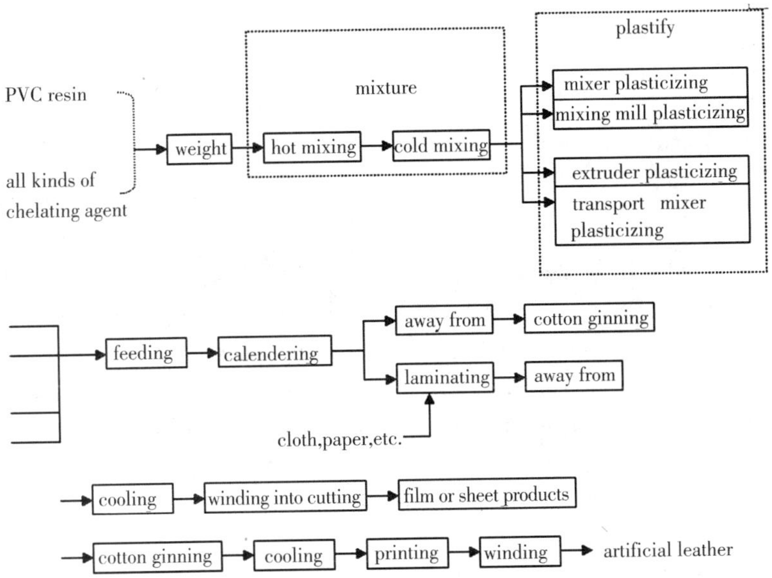

Fig. 4 – 8 Rolling forming process diagram of PVC calendering products

1. Soft PVC Film Calendering Process

Fig. 4 – 9 is the schematic diagram of production process of the soft PVC calendered film. Firstly, after weighing polyvinyl chloride and various complex agents according to the formula, add them into high speed mixer, and stir the material in high speed for a certain period of time at a certain temperature, then put them in cold mixer, stirring slowly and cooling to 60 ℃ below to prevent caking. Mixed material can adopt double roller mixer, mixer, extruder or mixing and transporting machine for plasticizing; molten material after plasticization can adopt extruder or two sets of double roller mill to feed the calender machine continuously. Feeding should be uniform, and the material temperature should also be relatively stable.

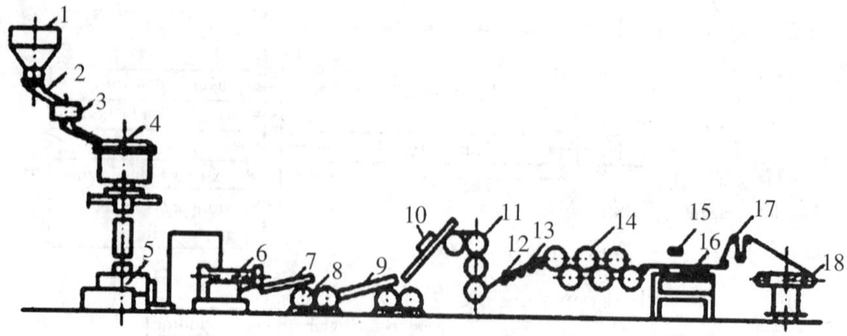

Fig. 4 – 9 Soft PVC film calendaring process diagram

(1—resin silo;2—electromagnetic vibrating feeder;3—weighing device;4—high speed hot mixer;5—low speed cold mixer;6—plastic extrusion machine;7,9—conveyor belt;8—two-roll mill;10—metal detectors;11—four-roll calender;12—traction rollers;13—roller;14—cooling roller;15—thickness gauge;16—conveyor belts;17—tension device;18—center winder)

Material from the feeding stage to calender should ensure the plasticizing fully. It must be uniformly mixed without impurities. Through the calender rolling it becomes a thin film of uniform thickness, then by the bearing of traction roller goes away from the calender, and then be stretched further, and decreases the film thickness to the desired target. If the surface of products demands patterns, then there should be cotton processing. Then after cooling, thickness, section and transport achieve the finished film by winding device volume. Thickness of film often uses β-ray thickness gauge to monitor continuously and control the feedback.

β 射线测厚仪

The difference between the sheet and film is mainly the thickness. Using 0. 3 mm as the dividing line, the thin is film, and the thick is sheet. They're both in the same production process.

2. PVC Leather Production Process

PVC leather is the main variety of artificial leather, which is made basing on cloth or paper and covered with PVC film. Its production method mainly has the coating and calendaring.

Rolling process before the laminating process of each working procedure in the production of PVC artificial leather production process is the same to the film rolling production process. In production, press PVC into required films of width and thickness on the calender, and stick together with the preheating of the duct through the calender roller extrusion and heat lamination, then undergo through the cotton ginning, cooling, slicing and coiling process. Finally PVC leather is made. This method is called method of direct attachment . In addition, there is the distributional layer conjunction i. e. after the calendered film forming into volumes,

直接贴合法
分布层合法

film and cloth are attached in the special joint production line.

Joint operation method includes erasing and rubberizing method, as shown in Fig. 4 – 10. Laminating method which has certain ratio between two doubling roller calenders is called erasing method. Roller speed in general to send plastic film is about 40% larger, so that partial plasticized plastics can be rubbed into the cloth, and can make the fit between the film and the substrate strong. While attaching method whose speed between two doubling rollers of calender is the same, is called adhesive method. There are inside and outside implementation ways. Inside implementation refers to the way that films and substrates are not fitting between two rollers of rolling machine, but to equip another doubling roller in the calender of a roller side, fitting between the film and the substrate is in the external fitting roller and roller through the appropriate pressure. Outside implementation way refers to after calendered film is away from the calender roller, also setting up a group of specialized laminating roller in the roller before the film and fabric lamination. When we use glue method, we often coat with a layer of glue on contact surface between film and the substrate to improve bonding strength. 贴合操作法

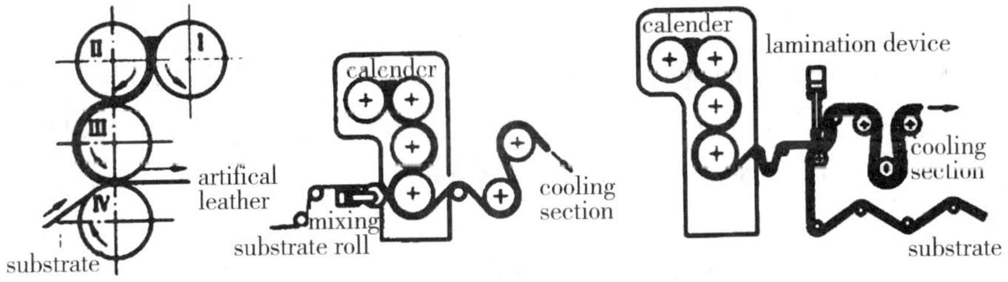

(a) brush glue method (b) indoor paste (c) outdoor paste

Fig. 4 – 10 Four roll calendar production of artificial leather schematic

4.3.2 The Calendering Mold Process Factors

Plastic calendering mold mainly includes the material factors and the rolling factors. The former includes material composition formulation, mixed and plasticized; the latter includes roller temperature, roll speed, roll speed ratio, roll spacing, roll clearance stock, cool forming, coiling and away from. 辊速比

1. Material Factors

(1) The Material Composition and Formulation.

聚合物/相对分子量

In general, if we use polymer with high relative molecular weight and narrow distribution of relative molecular weight, it is helpful to improve the product's physical and mechanical properties, thermal stability and surface uniformity, but this requires a higher rolling temperature and is not conducive to the production of thin products. Therefore the product properties and forming technology should be taken into account when selecting grade polymer resin. In addition, ash content, moisture and volatile chapter influence transparency and quality of resin film.

透明度

增塑剂

In a certain range, the higher plasticizer content is, the lower material viscosity is, the better performance of forming process is. Plasticizer varieties and amount will affect the heat resistance and optical properties of the product, so we should pay special attention to this when we make selection. If we use bad compatibility with polymer resin or high molecular polar groups electropositive stabilizer system, it is often easy to precipitate in the material surface during rolling, and then on the surface of the roller waxy layer adhering and forming, make the product surface flat, and there will be sticky roller in film production. Choose appropriate lubricant can overcome this problem to some extent.

(2) Mixed and Plasticized Material.

The material components should be fully mixed and dispersed, and plasticized evenly. If dispersion is uneven, product often appears fisheye, reduced flexibility, and other quality defects; and uneven plasticization can make products appear scar.

The plasticizing temperature should not be too high, the plasticization time should not be too long, otherwise it may cause the excessive plasticizer volatilization, then it is easy to cause decomposition of polymers. If plastication temperature is too low, inequality plasticizing will appear and calendaring can not be carried out because of material' s not sticky to roller. Plasticizing temperature of PVC depends on the specific formulation, generally in 150 ℃ - 180 ℃ range.

2. Rolling Factors

(1) Roller Temperature.

One of the heat material needed when installing calendering mold is heating device of calender roller. The other is the friction between the material and roller, but also the friction from internal shear of the material. Friction size, is related not only to the roll speed and roll speed ratio, but also to the viscosity of the material itself, which is related to the degree of increasing of sleeve temperature and plastic material. Therefore, in determining the calender roll temperature, we should consider effects of material formula and roll speed.

Too high roller temperature will cause too high material temperature, and tacky material bag roll, therefor rolling cannot be carried out; too low temperature, then the rolled products surface will appear roughness, opaque, bubble, and even the emergence of empty because of low temperature material.

In calendering mold, the plastic material has tendency of adhesion at high temperature and high speed roller, in order to make material in the roller transfer and prevent the air from getting into in the process of rolling, for the four-roll calender, roller temperature should be controled as $T_{roller III} > T_{roller IV} > T_{roller II} > T_{roller I}$. The $T_{roller III} > T_{roller IV}$ is designed to benefit film away without encasing roll. In general, temperature difference between rollers should be controled in the range of 5℃ - 10℃.

(2) The Roll Speed and Speed Ratio.

Suitable calender roll speed mainly depends on the thickness of material and products. Typically in calender roll speed, soft products ratio is faster than rigid products. The linear speed ratio of adjacent calender rolls is called the roller speed ratio. The purpose of demanding speed ratio between calender rolls is to make rolling material transfer smoothly in the calendering drum, also conducive of the material for further shearing plasticizing and calendering fabric for making extension and orientation of certain thin gradually. For the four roll calender, roll speed should be controled as: $v_{roller III} \geqslant v_{roller IV} \geqslant v_{roller II} > v_{roller I}$.

Roller speed ratio should be adjusted according to the thickness of the products and the roller speed, four roll calender in general (1: 1.05) to (1: 1.25). Three roll calender, roller speed ratio is generally 1: 1.05. In general, lower roller is constant, only for the ironing effect. Too high roller speed ratio will appear wrapping roller, while roller speed ratio is too small, soft products are not easy to absorb the roller, so that the air is easily clamped and this makes the product appear bubble. For rigid products, too small roller speed ratio will appear "cracking" 脱壳 phenomenon, and the plasticizing effect is not good, resulting in lower quality.

(3) Roll Spacing and Roll Clearance Stock.

In calendering, spacing adjustments between the rollers, one is to adapt to the different thickness requirements of products, and the other one is to adjust the inventory between roll clearance. In order to ensure that there is always some storage quantity in the backward roll gap, at the same time according to the rolling principle, pressure the material received between the two rollers increases with roller spacing decreases, so along the forward direction of the material, each roller spacing should be reduced gradually. Roller spacing decreasing gradually will increase pressure on the material gradually, in favor of improving the product density, eliminating foam and making further plasticization of material. The last pair of roller of calender roller spacing should be controlled to the same thickness of the product, but considering the subsequent processes of traction and that ginning will make the product thickness decrease, it should be reserved for allowance.

In calender roll gap, there should be a small amount of storage. Rolling stock is to guarantee the stable pressure of rolling process, to reserve and supplement material and further to improve the plasticizing and mixing dispersion effect. If the stock is too much, film surface will have a rough batocera phenomenon and blistering, and the surface of hard sheet stock surface will appear cold spots and uneven quality problems due to excessive cooling. But if the stock is too small, surface of product will be lightless because of the insufficient rolling pressure and fluctuation. Plate will appear rhombic hole. More serious product edge defect will be caused. At the same time deposit material should be in a pencil shape, and keep the rotating motion. If material and roller temperature is too low, or roll gap adjustment is undeserved, it can appear poor stock rotation, and thus lead to unevenness of transverse thickness of product, thin film bubble, plate with cold scar and so on. We should observe and adjust storage condition in production.

云斑现象

菱形孔洞

(4) Away from, Cool Forming and Coiling.

Products are led out by calender roller, through from the roller, embossing roller, the cooling roller and roller and then become products. In the follow-up process, in order to make flat-rolled products produce the necessary tension, the back-road roller speed should increase more successively than the front-roller, and there should be a certain ratio. Ratio speed between calender roller and roller away is relatively large, generally (1∶1.10) - (1∶1.35). The purpose is to make rolling products have certain extension and orientation, for the film emphasizes the unidirectional strength, and the ratio can be larger value. The rest ratio between the rollers is relatively small. Extension of product should be avoided after cooling, otherwise it will cause the internal stress of product, resulting in the increase of shrinkage rate. Film is not easy to expand, and sheet is easy to warp.

冷却辊

When cooling, cooling flow channel structure of the chill roll should ensure uniformity temperature of the cooling roll surface, otherwise it will lead to inconsistent shrinkage due to uneven cooling, film wrinkling or uneven after rolls and shrinkage rate will also be large. If cooling exceedingly, roller surface will have condensation of water droplets due to excessive low temperature. After products are coated with water droplets, it will be moldy or frosty during storage. We should pay attention to this particularly in humid summer.

4.3.3 Calendering Mold Process

1. Irregularity Product Thickness Problem

The most prominent quality problem of rolling products is that the uneven thickness of lateral film. The cause of transverse unevenness thickness of film is roller elastic bending deformation and the roller surface axial temperature difference. 弯曲变形

(1) Elastic Bending Deformation of the Roller.

In the rolling, the material gets pressing force by the roll, and the material bounds to generate the reaction towards roll, which is separation force. This separation force will make the calender rolls like the beam which will produce the elastic deformation; deformation value will be the biggest in the central part of the roll tube, and then decreases gradually to the ends, resulting in transverse cross-section of rolled products showing an intermediate and thin on the ends.

When calendering, total pressure material subjected and total separation force roller subjected are equal, so the separation force of roller can be obtained by the material stress p integral in the vise district, its value is:

$$F = 2\eta vR\omega\left(\frac{1}{h_0} - \frac{1}{H}\right) \qquad (4-7)$$

In the formula, F is the separation force; ω is the axial surface width of roller; H is the material thickness of the roll stock; η, v, R, and h_0 are the same as those in formula (4 - 1).

We can see that value of the separation force is associated with roller radius, speed, material thickness, viscosity, storage products, thickness, width and other factors. The higher the speed of roller is, the thinner and wider, the product is then roll separation force is more big. At this time, we can reduce the roll separation force through other adjustments, to control the uneven product thickness, such as increasing the calender roll temperature, reducing material viscosity or reducing the inventory. In particular, we adopt the height method (or bump coefficient method), roller shaft cross method or roller pre-bending method and other corrective 凹凸系数法

measures to compensate the problem of uneven thickness products led by elastic deformation of roll, as shown in Fig. 4 – 11.

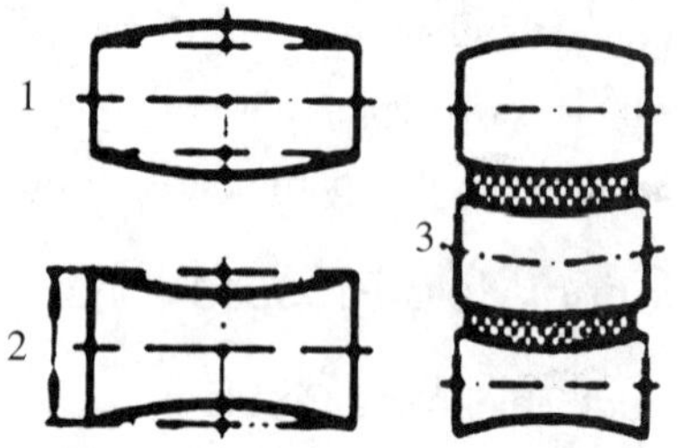

(a) roller bump coefficient method (or height method)

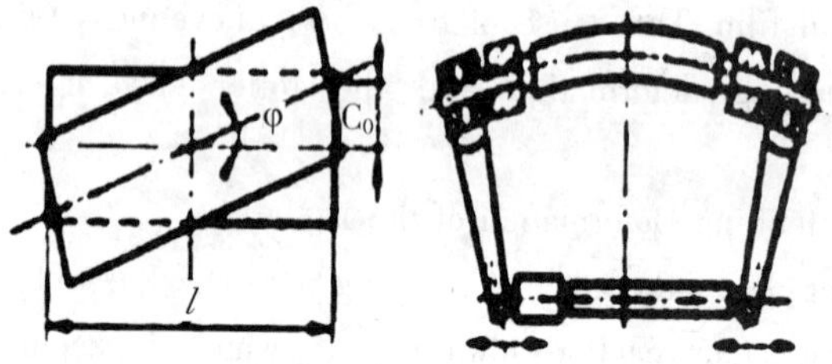

(b) roller shaft cross method (c) roller pre-bending method

Fig. 4 – 11 Calender roller deflection compensation method

(1—convex roller; 2—concave roller; 3—tree-roll calendar roller)

(2) Nonuniformity of Roller Temperature.

Because the bearing lubricating oil carries off heat and continues to transfer heat to roller frame, temperature of the ends of calender roller is lower than that in the middle. Roller temperature is not uniform, then it will inevitably lead to the uneven thermal expansion of the whole roll, and when cooling the product rolled out, shrinkage in the middle is high, and at both ends is small, so there will be products of thin in the middle and thick at both ends.

红外线

In order to overcome the problem of uneven transverse thickness resulted by different temperature of the roller surface, in process, we can use infrared or other methods to compensate heating in low temperature area, or near middle area on both sides of the roller, in order to improve uneven temperature of roller, we use the wind pipe cooling. However, key to ensure uniform transverse thickness of rolled products is height method, shaft cross method and method of pre-bending device.

2. Rolling Effect

Rolling product quality problem is the problem of calendering effect. Rolling effect is affected by temperature calender rolls, roll speed, ratio, roll away, storage quantity and nature of the material and other factors. These factors make rolling effect more prominent. The roll speed and roll speed ratio increase, then rolling effect improves; the material temperature increases, it is conducive to the solution of orientation, rolling effect decreases; roller spacing decreases, thickness of products becomes small, then shear effect of material increases, and the corresponding calendering effect improves; if the inventory increases, rolling effect increases. Acicular or flake compounding agent in material is easy to generate orientation effect, to reduce or eliminate the calendering effect generated by such factors. The main way is to use less or no such accessory ingredient with the worse shape symmetry. 配合剂

Chapter 5
Hollow Molding

中空吹塑

5.1 Outline

概述

吹塑模塑
空心塑料制品

Hollow molding (also known as blow molding) is a forming method of manufacturing hollow plastic products. It draws lessons from the glass blowing process with a long history, becoming plastic blow molding technology in 1930s. So far it has become one of the main plastic molding methods, and also has the great development in the blow molding method and molding machinery type.

型坯

Hollow molding is the process that makes hot melt plastic parison close in mold bulge into hollow products through gas pressure. According to the production characteristics of parison, they are divided into two kinds: ①Extrusion parison, squeezing tubular parison into two petal mold opening. When the parison reaches a predetermined length, the mold is closed, and cut off the parison, close the top and bottom of the parison, at the same time insert the compressed air into the tube center or the needles of parison wall, blow parisons close to the cavity, then extrude the products from the mold after cooling. ②Injected parison, injecting into parison with bottom in the mold with the method of injection molding. Open the mold and then move the parison to the blow mold for blow molding, and extrude products after cooling.

汽车工业
工程吹塑

Hollow molding products include plastic bottles, containers and hollow products of various shapes. They have been widely used in chemical industry, transportation, agriculture, food, beverages, cosmetics, medicines, cleaning products, children's toys and other fields. For the complex shape and unique features of office supplies, household appliances, furniture, culture and entertainment supplies and spare parts for the automotive industry, such as bumpers, gasoline and fuel oil, with the higher technical content and function, also are known as the "engineering blow".

Blow molding products possess properties of excellent resistance toward environmental stress cracking, air tightness (to prevent oxygen, carbon dioxide, nitrogen and water vapor from scattering to the inside and outside of the container), impact resistance which can protect the container, resistance to chemical reagents, antistatic property, toughness and resistance to extrusion. The commonly used hollow molding plastic includes polyethylene, PVC, polystyrene, polypropylene, ethylene-vinyl acetate copolymer, polyethylene terephthalate (PET), polycarbonate, polyamide and so on, among which polyethylene is the most widely used. Plastics whose melting index is in the 0.04 to 1.12 range is the better blow molding materials, used in the manufacture of various pharmaceutical packaging container. Low density polyethylene is used as food packaging containers; high and low density polyethylene mixture is used in the manufacture of various products container. Ultra high molecular weight polyethylene is used in the manufacture of large containers and fuel tank. Polyvinyl chloride plastics with excellent transparency and tightness is used for the manufacture of mineral water and detergent bottles. Due to its poor tightness and resistance to impact strength compared to PVC and polyethylene, polypropylene is not used widely in blow molding. Since using two-way stretch blow molding process, transparency and impact strength of PP are greatly improved, being suitable for producing thin-walled bottle, used for packaging container of detergents, pharmaceuticals and cosmetics. While due to its good transparency, high toughness and non-toxic, polyethylene terephthalate is largely used in bottles and so on. "The blow molding of plastic engineering" has been extended to ultra high molecular weight and high density polyethylene, polyamide plastic and alloy, POM, polycarbonate.

气密性

耐药品性/抗静电型

聚对苯二甲酸乙二酯

Hollow molding includes extrusion blow molding, injection blow molding and stretch blow molding. Stretch blow molding includes extrusion-stretching-blow molding and injection-stretch-blow molding. Its production process is composed of manufacturing of parison and inflation of parison. Difference between extrusion blow molding and injection molding is that the former is extruded manufacturing blank, and the latter is injection molding parison. Stretch blow molding has a longitudinal tensile bar, making the products in blow molding with transverse blown (tensile), and also in the longitudinal direction being stretched to improve its performance. Blow molding process is basically working as the same. Because the extrusion and injection molding are discussed, this section focuses on the characteristics and requirements of parison and blowing device, and their impact on technology.

挤出—拉伸—吹塑/注射—拉伸—吹塑

吹胀设备

挤出吹塑

5.2 Extrusion Blow Molding

挤出吹塑工业过程

5.2.1 Extrusion Blow Molding Craft Process

Extrusion blow molding process includes: ① extruding parison; ② when parison reaches a predetermined length, clamping after gripping the parison position; ③parison head forming and sizing; ④compressing air into the parison for blow, making it close to the mold cavity to form products; ⑤cooling the product stereotypes in the mold; ⑥extruding products from the mold, trimming, finishing on the product. There are many methods and types to achieve the process, which can realize the full automation operation. Extrusion blow molding methods and types are shown in Table 5 – 1.

Table 5 – 1 The way and the types of extrusion blow molding

The way of extrusion parison	Form of blow mold
Intermittent extrusion parison	A pair of mold or more pairs of mold to the parison
Continuous extrusion rotation discharging	Two vices—single cavity mold
	Several dies—multiple cavity
Continuous extrusion—delivers the parison	Horizontal delivery, vertical delivery
Continuous extrusion—with the reservoir	Large mold
Continuous extrusion—using screw or plunger pusher	—
Continuous extrusion—the moving mold	Move vertically, horizontally, rotary table
Continuous extrusion—refrigeration parison	

间断挤出/连续挤出

Extruded parison contains intermittent extrusion and continuous extrusion. Intermittent extrusion after parison reaches to the required length, and the extruder screw stops rotating and discharging, after the parison inflation finalizes the design to complete a production cycle (intermittent extrusion of parison, mold, blowing, cooling, sintering are carried out under the nose), then restart the extruder to extrude next parison. Due to the flow interruption of intermittent extrusion material, it is prone to appear thermal decomposition, while the extruder capacity can not play fully. It is used for polyolefin and non heat-sensitive plastic blow molding.

The continuous extrusion is continuous production of predetermined length parison by extrusion machine where the parison is received by the mobile mold, and cut off on the head, and sent to the blowing position or sent to subsequent process by transfer mechanism gripping the parison (molding of parison and the former type inflation, cooling, sintering are synchronized). Continuous extrusion parison includes reciprocating type, rotation discharging type and rotary type, applicable to a variety of thermal plastic resin molding. Melt plastic is less likely to thermal degrade, and can be applied to blow molding of PVC and other heat-sensitive plastic. Because continuous extrusion method can give full play to the ability of the extruder and improve production efficiency, it is widely used.

轮换出料式

热敏性塑料

5.2.2 Equipment of Extrusion Blow Molding

挤出吹塑设备

1. Extruder

The extruder should possess a driving device which can adjust speed continuously to extrude parison in a steady speed. Parison extrusion rate and the best molding cycle are coordinated. The draw ratio of extruder screw should be appropriate. If the ratio of length to diameter is too small, plasticizing of toughening material will be uneven; feeding ability is poor, and the parison temperature is not uniform. If the ratio of length to diameter is bigger, transition of heat and energy to material can be more uniform. The temperature fluctuation is small, and the heating temperature of cylinder is low, so that the parison temperature is uniform, it can improve the precision and uniformity of the parison, and is suitable for production of heat-sensitive plastics. For a given material storage temperature, low sleeve temperature can prevent the material from thermal decomposition. When the parison is extruded at low temperature, the melt viscosity is higher, which can reduce the parison sagging and assure uniform thickness of parison, and then shorten the production cycle and improve production efficiency. But the shear and pressure are higher in the extruder, so it requires transmission of extruder and the thrust bearing should be strong and durable.

2. Handpiece

机头

The handpiece comprises porous plate, screen connecting pipe, core component and so on. Design requirements for the handpiece are that: the flow should be streamlined; channel surface should have higher smooth degree, and there are no block parts to prevent from overheating decomposing in the nose due to unsmooth melt flow.

转角机头
带储料缸式机头

Blow molding handpiece is generally divided into three types: corner handpiece, straight through type nose and handpiece with a storage cylinder.

(1) Corner Handpiece.

It is connected by a pipe and a pipe head rectangular configuration composition. The structure is as shown in Fig. 5 - 1. This type of nose channel has higher compression ratio; die part has longer shaping segment, suitable for extrusion of polyethylene, polypropylene, polycarbonate, ABS and other plastic. Aiming at the defects, there is an improved form, as shown in Fig. 5 - 2 and Fig. 5 - 3.

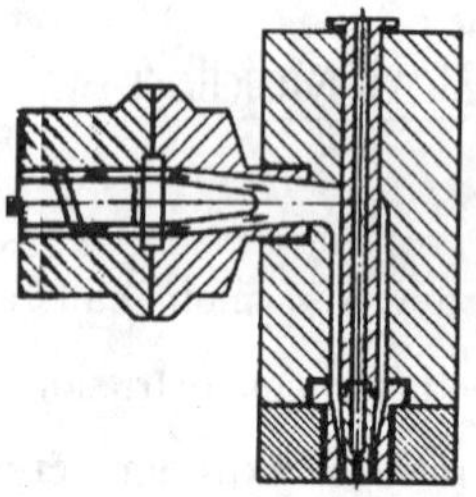

Fig. 5 - 1 The tube-type handpiece at right angles

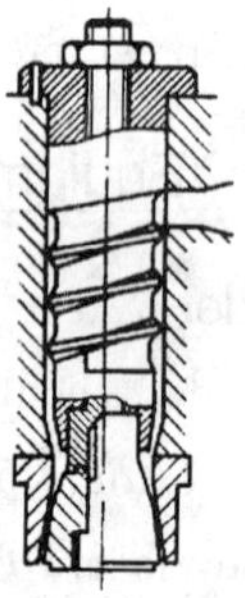

Fig. 5 - 2 The handpiece using helical to the parison extrusion direction groove mandrel

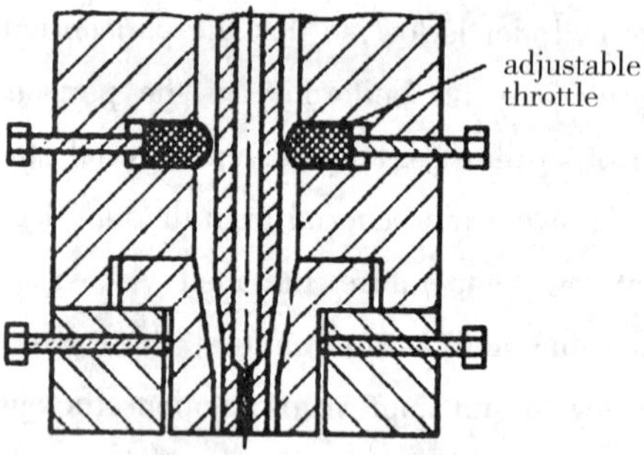

Fig. 5 - 3 Adjustable shift throttle-type handpiece

(2) Straight Through Type Nose.

Straight through type nose and extruder are in a glyph configuration, so as to avoid changing of the plastic melt flow direction, and prevent the plastic melt overheating and decomposing. Straight through type nose's structure can meet the blow molding of heat-sensitive plastic, commonly used in the manufacture of transparent bottle of hard PVC.

(3) Handpiece with a Storage Cylinder.

Producing large blow molding products, such as beer sleeve and the garbage box, because of the larger product volume, it needs certain wall thickness to achieve the necessary stiffness, therefore large parison needs to be extruded, but the tenesmus and decrease of large parison are serious; cooling time of product is long, and the requirement of output of extruder is lager. For large products, on one hand, they require quick provision of large amounts of melt quickly, reducing the parison drop and necking; on the other hand, if the large product's cooling period is long, the extruder can not run continuously, thus it has developed head with material storage cylinder. Its structure is as shown in Fig. 5 – 4.

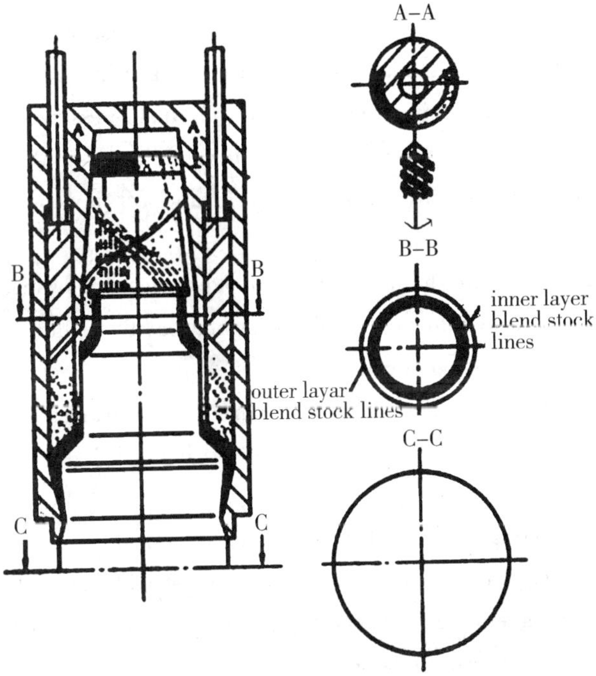

Fig. 5 – 4 Typical storage tank

By providing uniform plasticizing melt into the storage cylinder with extruder, the required amount of melt is stored in the storage cylinder according to a certain period. In the material storage system, timing is by the plunger cylinder (or screw). Stored material (melt) is launched intermittently and quickly. The large parison is formatted. Launching materials in high speed can reduce the large parison drop and necking, overcome the dropping and deformation of parison, and then avoid the inconsistent wall thickness of products. At the same time extruder can keep continuous operation, preparing for the next parison material. The machine can not only play as an extrusion machine, but also improve the speed of extrusion and shorten the molding cycle. But we should notice that when the plunger drives too fast, melt velocity through the head will be too large, which may produce melt fracture phenomenon.

In order to make extruder produce the parison of needed diameter, wall thickness and viscosity, we should reasonably determine the diameter of die, the width of die gap and the length of setting section. In determining the die diameter, first we should select the blow ratio which is suitable for product's outside diameter (the ratio of product's diameter and the outer diameter of the parison), and determine the maximum of outside parison. We should also consider the problem of die swell, and finally determine the diameter of die. If die gap's width is large, and the shear rate of resin melt is small, it is not easy to cause surface roughness as the breakage of parison melt. Setting section is long, and melt pressure of internal head rises, which is conducive to the elimination of weld lines, but the pressure loss appears. Case-hardened length(l) and the gap(t) ratio(l/t) is about 10 for fit.

3. Blowing Device

After parison gets into the mold and the mold is closed, blowing device would blow the tubular parison into precise shape as the cavity needed, and then cool down, shape, and sinter out products.

Blowing device includes blowing mechanism, mold, exhaust system, cooling system, etc. Now the first two points are mentioned below.

吹气机构

(1) The Blowing Mechanism.

Selection of blowing mechanism should be based on equipment condition, product size, and thickness of product distribution. Air pressure should base on the principle of outlining a clear pattern of products by blown parison. Generally the blowing mechanism has needle blowing, top-blowing and bottom blowing.

针吹法

①Needle Blowing Method: As shown in Fig. 5 – 5, blowing needle is mounted at the half-height of mold which is cavity. When the mold is closed, needle goes out forward parison wall, compressing air and blowing parison through the needle, and then the blowing needle is retracted; the molten material close to the pinhole is left behind by blowing needle. The other way is by the elongation in neck part, so that the blowing needle can insert easily, without damaging the bottleneck. In the same parison several blowing needles can be used at the same time, in order to improve the effect of blowing.

Advantages: Suitable for rotation blow molding of continuous production without cutting parison; small containers of the blow neck end to end, for the blow molding without neck; parison cutter can be equipped into the mold; more suitable for blowing the container with a handle, which itself close products which are not connected with the body.

Disadvantages: Because the parison is clamped at both ends, in order to obtain the qualified bottles, we need finishing processing for opening products. Mold design is more complex, not suitable for the blow of large containers.

② Top-blowing Method: As shown in Fig. 5 – 6, top blowing method is blowing through the core. With die neck up, when the mold is closed, parison bottom clamps. When top opening, the compressed air inputs from the core, which gets into the opening parison directly and determines the neck inside diameter, cutting off the parison between the core and the top of mold. Advanced top-blowing core method is composed of two parts. One part is the bottleneck diameter, the other part is a rotary cutter that slides on the blowing core. After blowing gas, rotary tool slides down, and removes the excess material. 顶吹法

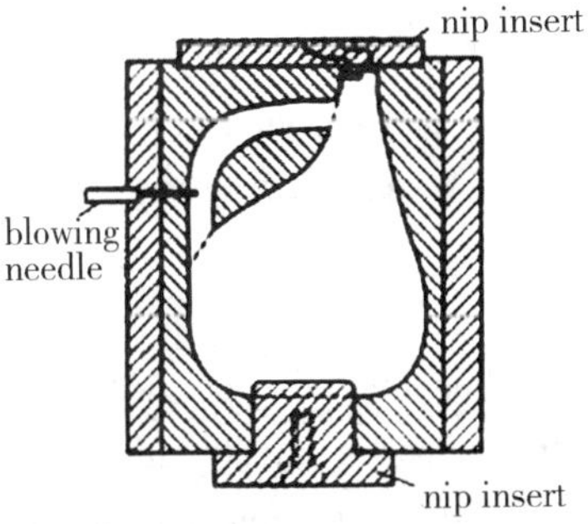

Fig. 5 – 5 Structure of blow needle

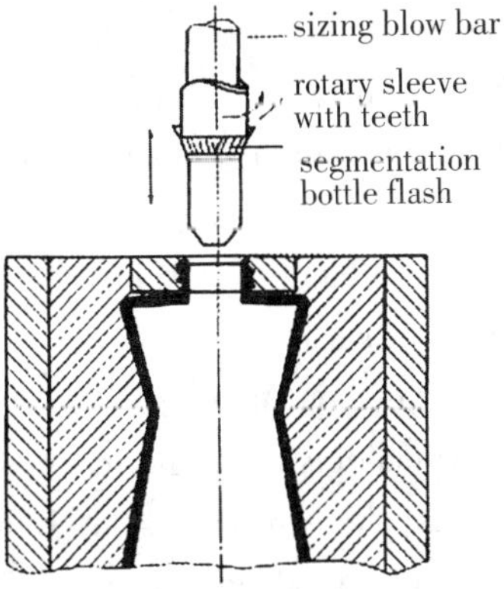

Fig. 5 – 6 Top-blowing device with the role of sizing and cutting diameter

Advantages: With direct use of core as blowing core shaft, compressed air is introduced from the crosshead above, getting into the parison through the core axis, which simplifies the blowing mechanism.

Disadvantage: Diameter and length cannot be determined, so an additional modification step is needed. Compressed air crosses from the core nose, to influence the nose temperature. Therefore, we should design independent top blown mandrel that has nothing to do with the die core.

底吹法

③ Bottom Blowing Method: As shown in Fig. 5 – 7. The extruded parison falls on the bottom of the mold core, inflating the parison through the core. The outer diameter of the core and die bottleneck coordinate with fixed internal and external dimensions of mold bottleneck. In order to ensure the accurate bottleneck size excess material should be provided in this area, which leads to the formation of two auricular flash in parting surface of the products on the bottleneck after molding, and needs following modification.

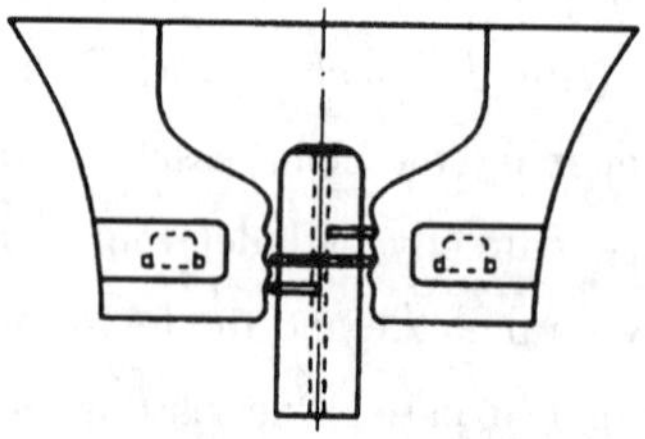

Fig. 5 – 7 Bottom blowing structure

Bottom blowing method is applicable to large vessels, in which the neck opening deviation from the center line of the products has irregular opening or a plurality of openings. Shortcomings of the bottom blowing method are that: gas inlet is in minimum temperature parts of parison, also is the thinnest section of the parison weight drop. When the product shape is more complex, it often cause insufficient inflation of product. In addition, clear traces leave after trimming the bottleneck of ear edge.

(2) The Blow Molding.

Blow molding is usually composed of two petals, and equipped with a coolant channel and the exhaust system. Blow mold structure is relatively simple, and the pressure under production is small. The mold strength requirement is not high, so aluminum, zinc alloy, beryllium copper and steel are often used. Mold cooling system directly influences the performance of products and production efficiency, so die has three sections to cool, adjusting the cooling water flow in every section according to the product shape and the actual needs, to ensure the quality of the products. The exhaust system of mold is used to discharge the air between blank and die cavity wall when blowing parison. If the exhaust system is poor, blowing gas in cavity will be forced to compress and trapped between the blank and the die cavity wall, so that the parison will not close to the cavity wall, resulting in that sag and wrinkle on product surface, pattern and the handwriting are not clear. They will not only affect the appearance, but also reduce the strength of the product. Therefore, the mold should set the exhaust hole or exhaust groove.

4. Auxiliary Device

(1) The Parison Thickness Control Device.

When parison is extruded from the die, expansion appears, then the parison's diameter and wall thickness will be greater than the die gap. Parison hanging in the die sags due to gravity, causes elongation, the longitudinal uneven thickness and wall thinning (of extrusion end wall thickness thinning) and effects parison dimensions and the quality of the products. Control methods of parison dimensions are: to adjust the die gap, change the extrusion speed, make a blow in advance, change the parison traction speed, and do parison thickness control.

The wall thickness of blow molding products depends on parison swell ratio of each part. On the same parison, the larger the swell ratio is, the thinner the wall will be. The smaller the blowing ratio is, the thicker the wall is.

For hollow products with complex shape, in order to obtain the product of uniform wall thickness, different parts of cross section of the wall of the parison thickness should vary according to the size of the blowing ratio. Modern extrusion blow molding parison program control unit is based on the requirements of uniform wall thickness of product. It determines the parison blow ratio of various parts of cross-section along the length direction, drawing the parison program curve through the computer system, controlling the thickness of cross section through the variation of control system. Thickness of cross section of parison along the length direction varies. The more varied parts (i. e. more points), product wall thickness will be more uniform. Determining parison wall thickness variation of cross section of program points should accord with the parison inflation ratio, which is called "green program". Distribution program points can be linear or nonlinear. The more program points, the more uniform of wall thickness is, and the more raw material can be saved. In these five parison thick methods, the widely used way of adjusting the die gap is shown in Fig. 5 – 8.

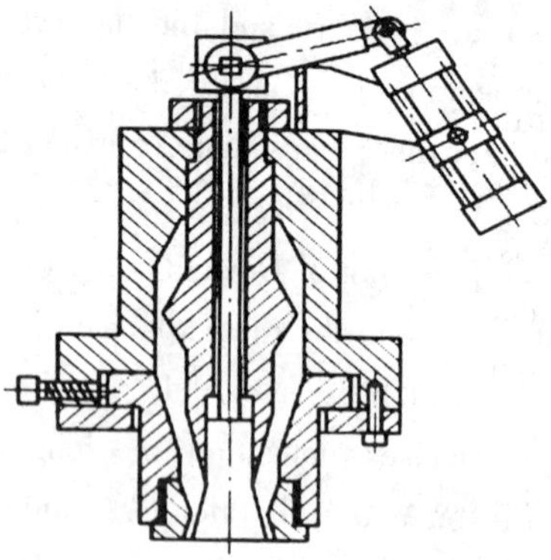

Fig. 5 – 8　Cone-die controlling parison thickness

(2) Parison Length Control.

Length of parison directly affects the quality of final products from blow molding and the length of tails cut, which determines the amount of raw materials consumed. Parison's length depends on the rotating speed of the extruder screw within a blow molding cycle. The photoelectric control system is generally adopted to control the parison's length. Through the photoelectric tube, changes of the ratio of actual and preset lengths of parison can be detected, then the rotating speed of extruder screw can be automatically modified by the control system, thus to calibrate the parison's length and reduce the external influence on the length of parison. This system is simple and useful, raw-material-saving and can reduce 5% of tailings. Generally, the system to control the thickness and length of parison is applied cooperatively with the photoelectric control system.

光电控制系统

(3) The Parison Cutting Device.

型坯切断装置

The parison should be cut at the required length. And cutting device should be applicable to plastics with various properties. The blow mold is made in halves and both ends of its chamber are fashioned into knife-edged cutters to cut off parisons. The edge of cutter, which can directly affect the quality of products, falls into several categories, such as flat-edge cutter and triangle-edge cutter. The UPVC transparent parisons would usually be cut with a heated flat-edge cutter.

5.2.3 Control Factors of Extrusion Blow Molding

挤出吹塑控制因素

The influencing factors on the technology of extrusion blow molding and the quality of the hollow products mainly include: parison temperature and extrusion speed, air blowing pressure and gas drumming rate, blow-up ratio, mold temperature, cooling time, etc.

1. Parison Temperature and Extrusion Speed

The temperature of parison is a factor affecting directly the visual quality, longitudinal wall thickness, and production rate of hollow products. When the parison is extruded, the temperature of melt should be uniform and properly low to increase rigidity of the melt, thus avoiding parison being drawn down by its own weight and improving the cooling rate of products to elevate production efficiency.

When over-heated, the parison will be extruded at an undesirable slower rate and easily drawn down by its own weight. As a result, the even longitudinal wall thickness will fail. The cooling time will be longer and the melt will fail to form products for lack of hot strength. When under-heated, the parison will have an obvious die swell, shrinking length and increasing thickness, which can undermine its visual quality for flow marks and its evenness in wall thickness, and additionally make products fragile, rough and dull.

In extrusion blow molding, the curling-up of parison is an easy occurrence, a result from unevenness of radial thickness of parison, and it is always curling up to the thinner side. The nonuniform temperature of parison can lead to unevenness of wall thickness, therefore, a fine control over the temperature of parison is indispensable. And the production method for that end is to, within the capacity of extruder, keep the temperature properly low and stable and speed up rotation of the screw, thus enabling extrusion of a parison, of smooth surface, even thickness and less drawdown.

2. Air Blowing Pressure and Gas Drumming Rate

鼓吹速率

Inflation here refers to injecting the parison with the compressed air which can blow it up against the wall of mold chamber and also cool it down. Parisons, made of different types of plastics, at different temperature and with different modulus values, need different air pressure to blow up, usually ranging from 0.2 to 1 MPa. For parisons made of polycarbonate plastics which have a high viscosity and modulus values, the air pressure should be set at the higher end of that range. By contrast, for parisons made of polyamide plastics which have a low viscosity and high ductility, the air pressure should be set at the lower end of the range. In other conditions, the air pressure should be set in the middle of the range. Blow air pressure and parison wall thickness, the volume size of products are related; thick-wall small-volume products can be used for low pressure blowing. Because the thickness of the parison is big and slow cooling, the melt viscosity does not increase quickly which can hinder it to swell; for thin and large volume products, it needs to adopt the higher blowing pressure to ensure product integrity.

Blow rate refers to the rate of the air volume filled. The high blow rate can shorten the parison's swell time, making the products of uniform thickness and good surface quality. But if the blow rate was too large, it will have a partial vacuum in the air inlet, causing the parison's invagination of this part, and even generating partial vacuum in the air inlet part. For this, we need to increase the air blowpipe diameter. When blown flasks can not increase the blowpipe caliber, we can only reduce the volume rate.

吹胀比

3. The Blow-up Ratio

Blow-up ratio refers to times of the parison and makes it as large as itself after blowing up. For parisons of the same size and quality, the greater the blow-up ratio is, the greater the size of the product will be. If the blow-up ratio increased inappropriately, wall thickness of products will become thinner. Blowing up will become difficult to handle although less raw materials are consumed, and the strength and rigidity of products will decline; if the blow-up ratio is improperly low, the raw material consumption will increase; the product thickness and the available volume will decrease, and the cooling of products will last longer, then costs will become higher. Generally speaking, blow-up ratio ranges from 2 to 4, which should be set according to types and properties of plastics, required shapes and the sizes of products and the sizes of parisons. Usually, for the large and thin products, the blow-up is lower, ranging from 1. 2 to 1. 5; for small and thin products, the blow-up ratio is higher, from 2 to 4. In making narrow-mouthed bottles, the blowing ratio can be 5 to 7 times higher.

4. Mold Temperature

Mold temperature directly affects the quality of the products. Mold temperature should keep uniform, to ensure uniform cooling of products. If mold temperature is too low, the parison will cool fast, and become difficult to change shape and swell with decreasing extensibility in the clamped part which, as a result, will become too thick. Through increasing air pressure and rate of pep, though somewhat overcoming, it still affects the evenness and the thickness of the products. The products and decorative pattern are not clear; the outline of product surface even shows spots and orange peel. If the mold temperature is too high, the time of cooling will extend; the production cycle will increase. When cooling is not enough, the products' are easily to change the shape after sintering and the shrinkage rate are large.

Usually, the small thick products, mold temperature is low; for large thin-walled products mold temperature is higher. The mold temperature's level should be based on the plastic varieties. For engineering plastics, with high glass transition temperature, which can be released without affecting the product quality at a higher mold temperature, high mold temperature is helpful to improve the surface degree of finish products. Blow mold temperature is in general lower than the plastic softening temperature for which 40 ℃ is appropriate.

5. Cooling Time

Parison inflation should be carried out after cooling, whose time controls the appearance of the product quality, performance and production efficiency. Increasing the cooling time can prevent the deformation caused by the elastic recovery of plastic products, and produce regular shape, clear surface pattern, and good quality. However, while the toughness and transparency are lower, then the production cycle will extend, then the production efficiency may be low.

Usually, on the premise of guarantee of sufficient cooling we can finalize the design of the finished products; speed up the cooling rate, and improve production efficiency. The methods of increasing the cooling rate are to: increase the cooling area of mold; use chilled water or refrigerant gas cooling in the mold; use liquid nitrogen or carbon dioxide to swell and cool the parison. 液氮 二氧化碳

Mold cooling rate depends on the cooling mode, selection of cooling medium and the cooling time, and also is related to the temperature and the thickness of the parison. With increasing product wall thickness, cooling time will extend. Due to the different thermal conductivity and different varieties of plastic, the cooling time can vary. Under the same thickness, high density polyethylene needs a longer cooling time than polypropylene. For a large thick wall and special shaped products, balance cooling can be used. The necking and the cutting site select the cooling medium with high efficiency; body thin section selects the average cooling medium. 聚丙烯

The special products also need a second cooling, and use air-cooled or water-cooled after product release, making its sufficient cooling to prevent shrinkage and deformation.

To sum up, the advantages of the extrusion blow molding are that: ① it has applicability to a variety of plastic; ② the production efficiency is higher; ③the parison temperature is uniform, and the broken product is reduced; ④ it can produce large-scale container; ⑤ it has less investment in equipment and so on. Therefore, the extrusion blow molding is still dominant in the current hollow production.

注射吹塑

5.3 Injection Blow Molding

注射吹塑生产工艺

5.3.1 Injection Blow Molding Production Process

Injection molding is the two-step forming method for producing hollow plastic container, its production process is shown in Fig. 5 – 9. By the injection machine under high pressure, molten plastic gets into the tubular blank forming parison mold, mold after parison in core (also called mandrel). Through mechanical device, it will heat the preform in blow molding mold. The compressed air of 0.2 – 0.7 MPa is introduced by the mold core channel after molding, so that the parison is blown as the shape of the cavity, and cooled in air pressure, forming final products after sintering.

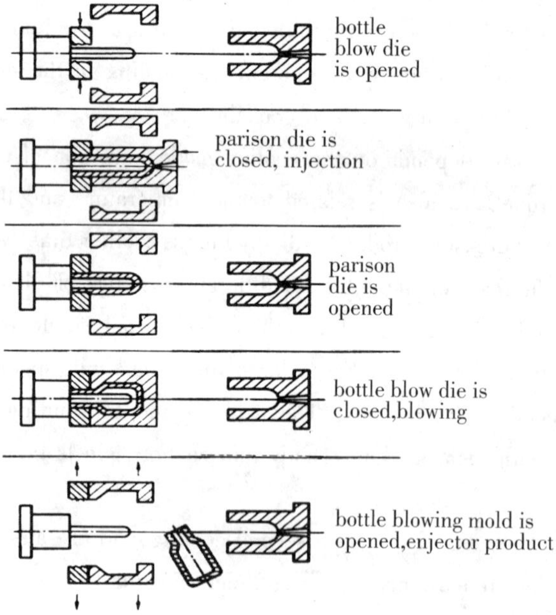

Fig. 5 – 9 Injection blow molding process

Injection blow molding is suitable for production of small refining container and a wide mouth bulk container. The maximum volume in general can produce no more than 4 L. Hollow containers by injection blow molding are mainly used in cosmetics, daily necessities, medicine and food packaging. The commonly used resins are PP, PE, PS, SAN, PVC, PC, etc.

Compared with extrusion blow molding, injection blow molding method has the following advantages. The products, with even wall thickness, do not need to be modified after processing; the products don't have the joint line, leading to less waste and scrap. The shortcomings are that: every product must use two kinds of molds (parison injection mold and blow molding mold); parison injection mold has to withstand high pressure; the two molds' positioning tolerance are with a high level; a higher mold cost is needed, the shape and size of production are restricted (it's hard to produce the container with a handle). Due to the above points, this method is still in development.

5.3.2 Characteristics of Injection Blow Molding Equipment

注射吹塑设备特点

The basic characteristics of injection blow molding parison are that: it's completed in injection molding; products are completed in a blow mold. Injection blow molding equipment is divided into two-position, three-position and four-position of the points. Design based on the above principle are called two machines (about 180° apart). Removal of products adopts mechanical hydraulic ejector mechanism to complete. Two-position machine has greater flexibility. Three-position machine is 120° apart, namely increasing special station removal products. Four-position machine is 90° apart, and is based on three-position machine's technical requirements for the additional station (preform is pre-blowing or pre-stretching). The most commonly used are these three kinds of machines, which account for more than 90%.

1. Requirements for Design of Injection Mold Cavity and Core

The injection mold cavity includes two halves mold, mandrel, bottom plate and collar, according to the shape of products, size, thickness, plastic shrinkage and the blowing of the overall design of parison shape. In addition to the neck of the bottle, it requires radial wall parison thickness greater than 1.5 mm, no more than 5 mm. The wall thickness should not be too thin to decrease the blow performance, or too thick to make it blow.

芯棒

Parison shape should be determined first, and then the shape of the mandrel should be designed. As the core rod should emerge from the container, the followings are required: ① mandrel diameter should be less than the minimum diameter of blow molding containers' neck, so that the plug can escape from containers; ② the minimum diameter of container is as large as possible, making the blowing ratio not too small, in order to ensure the quality of products. Mandrel structure diagram is shown in Fig. 5 – 10.

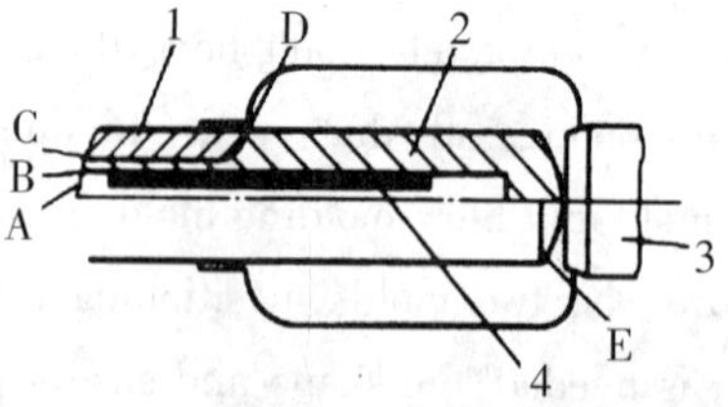

Fig. 5 – 10 Mandrel structure diagram

(1—slide holder; 2—slide; 3—bottom plug; 4—heating oil catheter; A—heated oil inlet; B—heated oil exports; C—airway; D—blow port; E—when L/D is large, bottle blowing port)

Mandrel has three functions. In the injection mold it reacts as the center to serve as the male mold molding parison; acting as the vehicles will be blank by injection mold delivery into the blow mold to a heating channel; the mandrel, using oil as the heating medium, controls the temperature; the mandrel has blowing channel; the compressed air can reach the bottom of the container and blow. A control switch device in the airway is opened so that a mandrel, blowing, injection molding is closed. Because of the functions, and being on the inside surface, formation of the container is so high and mandrel surface requires accuracy. Mandrel should have sufficient rigidity, good toughness, high surface hardness, corrosion resistance. Other parts' material requirements of higher mold, should be the oil quenching tool alloy steel and chrome production.

淬火

When the parison and the mandrel are determined, the injection mold cavity shape has been determined. To ensure the machining precision, the structure often uses a nest type, including the mandrel positioning injection cavity, cavity block, blowing mold plate and a screw thread mouth. In order to satisfy the requirement of heat transfer cavity block partition heating or cooling, the cavity cover and the cavity block should be selected over H_7/h_7. Mold should be used around the wedge position, and should not be column positioning guide. As the mold requires strength and high hardness compared with ordinary injection molding, high carbon steel or carbon tool steel cavity seat is used. The cavity covers the same material as

碳钢

mandrel.

2. Design Requirements of Blow Mold

Blow mold is the key device of container forming, showing the container shape, surface roughness and surface quality directly. Therefore, the mold should ensure adequate cooling to finalize the design in inflation, with the surface selection of tolerance limits, to prevent the seam line on products surface, in order to make the gas mold blowing process entrained rule out smoothly. It should be opened by several exhaust slot in the mold surface according to the shape of the container with the exhaust grooves of the depth of 15 – 20 μm, and 10 mm in width. The bottom of the container should be designed so that the sintering is concave. Generally, the soft plastic container bottom concave is 3 – 4 mm; hard plastic container bottom concave of 0.5 – 0.8 mm is enough. Special requirements can be designed to form the base with scalability. Phantom material generally uses carbon tool steel and ordinary steel corrosion in manufacture.

5.3.3 The Main Points of Injection Molding Process

注射吹塑工艺要点

1. Tubular Temperature and Blow Temperature

When injecting parison, tubular temperature is very important. Humidity too high and low viscosity calcined material is easy to deform. The tubular in the transfer of uneven thickness affects the quality of blow molding products. Temperature too low, often with more products within the internal stress, is prone to deform and stress rupture when using.

管坯

Choose the mold temperature as required, with configuration of mold oil temperature regulator, controlled by the high precision digital temperature controller (temperature range of 0 ℃ – 199 ℃). The temperature difference is between the < + 2℃. General configuration also has larger refrigerating capacity (23 KW) of water-cooled machine, which will help shorten the production cycle so as to save costs.

数字温控仪

水冷机

2. Injection Blow Molding Resin

The molecular weight and melt viscosity of resin for blow molding injection not only have higher melt viscosity, but also are influenced by the shear rate. The processing temperature for mineral products with good toughness, suitable melt elongation property, should ensure that the products' all edges and corners can uniformly present blow molding die cavity outline, and thickness will not appear noticeably thin or uneven thickness.

拉伸吹塑

5.4 Stretch Blow Molding

Stretch blow molding refers to the biaxial orientation stretch blow molding. It is based on common extrusion blow molding and injection molding. Extrusion or injection molding billet legal system first, and then the parison handle is put to plastic appropriate stretching temperature, the external (with drawing mandrel) or internal (tensile fixtures) mechanical force of the longitudinal tensile, at the same time or later is put by compressed air blowing for transverse tensile. Finally products come out.

非结晶型的热塑性塑料

The stretch purpose is to improve physical and mechanical properties of plastic materials. As for the amorphous thermoplastics, stretching is performed within the hot elasticity. For the fractional crystallization of thermoplastics, stretching process is in the temperature range below the crystalline melting point. During the stretching process, it must keep a certain tensile speed. Before its role is to blow molding, make plastic macromolecular chain stretching orientation instead of having direct relaxation.

晶核生成速率

成核剂

At the same time, you also need to consider the rate of crystal nucleus formation rate and crystal growth rate, when the crystal has not yet formed, even reached the proper stretching temperature. The parison is pointless, therefore, in some cases, nucleating agent can be added to increase the nucleation rate.

阻隔性

Beam and radial have the importance of orientation, excellent performance of container appearance, product transparency, impact strength, hardness and rigidity. Surface glossiness and barrier property have increased significantly, which is shown in Table 5 – 2.

Table 5 – 2 Property of parison directional injection stretch blow molding bottle

熔融指数

Molding method	Billet injection blow molding	Injection stretch blow molding parison orientation		
Material: polypropylene				
density/(g/cm^3)	0.9	0.9	0.9	0.9
melt index /(g/10 min)	7	8	7	7
Product properties				
volume/cm^3	500	180	450	1 350
quality/g	38	8	12	12

(续上表)

Molding method	Billet injection blow molding	Injection stretch blow molding parison orientation		
Tensile strength /MPa				
longitudinal	45	75	70	125
transverse	45	86	65	115
Elongation/%				
longitudinal	750	56	50	50
transverse	750	60	100	50
Young's modulus /MPa				
longitudinal	950	1 280	1 290	1 370
transverse	950	1 690	1 140	1 470
The degree of fog /%	50 - 60	5.2	13.3	10.5

拉伸强度

杨氏模量

雾度

5.4.1 Stretch Blow Molding Process

拉伸吹塑工艺

Stretch blow molding process is divided into the one-step method and two-step method.

One-step method refers to the three main parisons, stretching and blowing process continuous completed in a machine, also known as hot parison method. Parison is semi-finished products in the production process. Equipment combinations are as follows. ① It is composed of extruder and blow molding machine; ②It is composed of injection machine and blow molding machine.

In two-step method production, the first step in the preparation of parison, parison after cooling to be processed into a semi-finished product, has the characteristics of the specialized production. The second step will be cold parison with another enterprise or in a workshop for reheating, stretching and blowing, also known as the cold parison method. For two-step production, process condition control is an incomparable step, processing method is suitable for mass production, however, it requires more energy consumption.

At present there are four types of stretch blow molding combinations: ① One-step extrusion stretch blow molding, used for processing PVC; ② Two-step extrusion stretch blow molding, used for processing the PVC and PP; ③ One-step injection stretch blow molding, used for processing the PET and RPVC;

④ Two-step injection stretch blow molding, used for processing the PET.

Stretch blow molding process includes: injection parison directional stretch blow molding, extrusion parison directional stretch blow molding, multiple directional stretch blow molding, compression directional stretch blow molding, etc. The characteristic is to control the parison temperature below the melting point temperature using two-way stretch to improve the strength of the product. Here are two main types of the stretch blow molding process.

1. The Injection Parison Directional Stretch Blow Molding

The first injection is molding a parison, and continuously from the conveyor belt (or rotary belt) sent to a heating furnace (infrared or electric heating), by heating to stretching temperature, then into the blow mold by means of tensile bar and axial tension, finally by the blow molding. Injection stretch blow molding is shown in Fig. 5 – 11. Process characteristic of this method is to increase the tension rod, which will be blank in blow molding machine and is usually 1 to 2 times of the axial tension. Therefore the tensile temperature needs to be controlled. This method can be used in multiple cavity (2 – 8), and production has a capacity of 250 – 2 400 /h (capacity of 340 – 1 800 g beverage bottles).

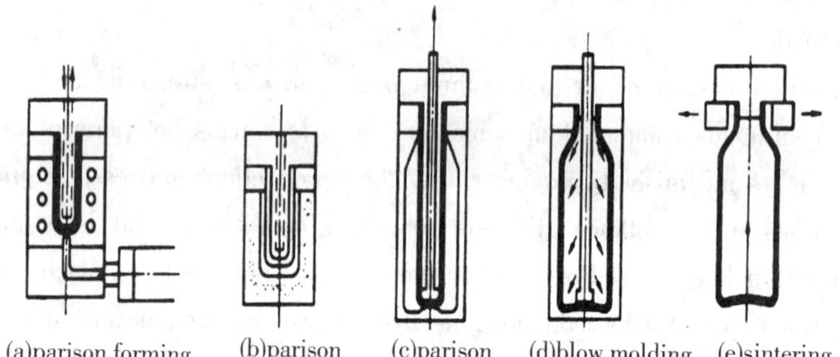

Fig. 5 – 11 Injection stretch blow molding

2. Extrusion Parison Directional Stretch Blow Molding

It is put in the heating furnace to the stretching temperature, and then through the conveyor will be heated. Parison is removed from the oven to the forming stage. The end of the parison formation bottleneck and thread, and the axial tension along the 100% – 200%, closed blow molds are blown. The other method is to remove from the oven parison heating, while the tensile device is along the tube axial tensile, side to blow mold, mold clamping the parison stretching after blow forming, trimming waste edge. This method of production capacity can reach 3,000/h; the capacity is 1 L. In order to meet the requirements of different processes, so far, it has developed a variety of industrial molding equipment, and has been developed multi-layer directional stretch blow molding.

5.4.2 The Key Points of Control of Stretch Blow Molding Process

拉伸吹塑工艺控制要点

1. The Selection of Raw Materials

In general, thermoplastic can stretch blow molding. By biaxial stretching, it can significantly improve the tensile strength, impact toughness, rigidity, transparency and gloss, improve the oxygen, carbon dioxide and water vapor barrier. From the current level of technology, plastics which can meet the requirements are: polyacrylonitrile (PAN), polyethylene terephthalate (PET), polyvinyl chloride (PVC) and polypropylene (PP). The maximum amount of use is polyethylene terephthalate, the minimum amount of use is polyacrylonitrile. Polypropylene's barrier to water vapor is good; after the biaxial oriented, the low temperature (5 ℃ – 20 ℃) brittleness is improved.

聚丙烯腈

脆性

2. Injected Parison Process Control

Temperature, pressure, time are the three major factors of parison injection. The melting temperature of sleeve temperature takes control of resin; injection pressure affects the product shape and precision; the molding cycle determines the production efficiency. Three factors restrict and influence each other. Take the PET resin bottle as an example. As PET resin as a crystalline polymer, crystallinity of 50% – 55%, has obvious melting point of 260 ℃, decomposition temperature greater than 290 ℃, then a sleeve nozzle temperature should be controlled between 260 ℃ – 290 ℃. The melt in viscous flow stays with lower viscosity, with moderate injection pressure and speed. Injection mold temperature is related to the performance, which directly affects the stretch blow molding process next. Injection of PET melt cavity must be in the T_g (70 ℃ – 78 ℃) following the rapid cooling, fast through the PET crystallization rate the fastest temperature (140 ℃ – 180 ℃) area to obtain transparent amorphous form parison. Therefore, the parison mold must use chilled water for cooling. The water temperature is 5 ℃ – 10 ℃, which will make the mold temperature drop to 20 ℃ – 50 ℃. The crystallization of PET parison is less than 3%. The molding cycle, temperature and freezing water run well, not only shortening the molding cycle, but also improving the product

粘流态

quality. The parison rapidly cools to below T_g, which can reduce fogging and crystallization. At the same time of resin retention time at high temperature, the acetaldehyde content is generated by thermal degradation of low.

Cooling time varies with the square of parison wall thickness. Parison wall is thick; the cooling time is long; the production cycle lasts longer. The parison is depending on the cooling water to take away the heat in the mold during cooling. When cooling water becomes turbulent in the channel, the heat transfers. Therefore, it should be equipped with high pressure and large flow pump. Injection molding process parameters PET performance is shown in Table 5 – 3.

水泵

Table 5 – 3 Injection molding process parameters

Cylinder temperature/℃	Injection pressure/MPa	Injection time/s	Temperature of the cavity/℃
In the . . . after 288 In the middle 282 In the. . . before 282	High pressure 76 Low pressure 41 Back pressure 0 – 35		1# 266 – 277 2# 266 – 277 3# 277 – 304
Screw plasticizing time/s	Cooling time/s	The cooling temperature/℃	4# 277 – 304
8.5	9	4.4 – 10	5# 277 – 304
Cavity number	Characteristics of mold	The amount of each cavity injection/s	6# 277 – 304 7# 277 – 304
8	Hot runner	68	8# 277 – 304

3. Process Control Characteristics of Stretch Blow Molding

Stretch blow molding is mainly affected by technological factors: parison heating temperature, molding pressure, molding and blow mold temperature. The introduction is as follows.

(1) Parison Reheat.

It is the characteristic of the two-step production. When the parison is taken out after cooling to room temperature, injection mold proceeds after 24 h of storage in order to reach thermal equilibrium. The parison reheating aims to increase the wall temperature to reach thermoplastic range, in order to stretch blow molding, obtain sufficient biaxial orientation, make PET bottle to reach the required physical properties, and make the goods transparent, high luster, without defect and the

光泽

uneven surface. Reheating is generally with far-infrared quartz heater or cooler. There is constant temperature box of linear, parison along the orbit transportation, fixed shaft to rotate the parison, moderate heat. Parison is out when the temperature is 195 ℃ - 230 ℃. It should be done in 10 - 20 s of the heat preservation, in order to achieve the balance of the temperature distribution.

(2) Stretch Blow Molding and Biaxial Orientation.

Parison after reheating should reach the temperature of the glass transition temperature 10 ℃ - 40 ℃, and then be sent to the stretch blow molding mold clamping. Axial longitudinal tensile rod moves up and down the stretch, the compressed air is piped in. In the circumferential direction of the parison transverse expansion, when stretch at the bottom of the stem to die. The parison swells after cooling. The tensile bar goes back, compressing air stops trending and obtains products.

Stretch ratio (including tensile rate, tensile length, blowing air pressure and rate) is the key factor to affect the quality of products. Longitudinal stretch ratio length and the ratio of the parison length (products) are implemented through the tensile bar. Lateral stretch ratio (the ratio of the parison diameter and the diameter of products) is achieved by blowing air. Practice has proved that when transverse stretching 5 times and longitudinal tensile 2.5 times, the quality of PET bottles is satisfied. The more times longitudinal and transverse tensile ratio are, the products will be bleached and strength decreased. 拉伸比

PET bottle blowing air pressure of the bottom bracket only needs 20 Pa. PET bottle without bottom support requires 30 Pa, then blow molding products get better quality. Blow air is prohibited in water and oil, or thermal expansion will produce steam, and the products formed on the inner wall of the bottle, in loss of transparency, affect the appearance and health. 热膨胀

In addition, there are multi-layer blow molding, large-scale hollow molding and so on, which are not going to be expanded herein. 多层吹塑

热成型

Chapter 6
Thermoforming

概述

6.1 Outline

Thermoforming is a method using thermoplastic sheet as raw material for the manufacture of plastic products. In manufacturing, the first step is to cut the material into sheet size and solid sample, to clamp on the frame and to heat it to the thermoelastic state, and then the applied pressure makes it close to the mold surface, thus it achieves the similar shape and profile. Sheet cooling after molding can be removed from the mold. After appropriate dressing, namely sheet becomes products. The applied pressure mainly depends on the pressure difference on both sides of the sheet, but also on mechanical pressure or liquid pressure.

热弹态

The characteristic of thermoforming products is thin wall, because the thickness of sheet as raw material is generally 1 – 3 mm (most are lower; a few special products with the thickness of sheet are thinner to 0.05 mm), and the product thickness is lower than this. If required, thermoforming surface area can be large (the sheet area with a single product is up to 3 m × 9 m), which is characteristic of this product. But thermoforming products belong to the half shell (concave and convex), have certain limit to its depth. Product types are cups, plates and other export utensils, medical electronic instrument accessories, radio and TV shell, billboards, bathtub, helmet, toys, packaging equipment and auto parts, building components, chemical equipment, radar cover and the aircraft cabin cover, etc. Choosing a variety of products of plastic varieties depends on the product and the use of performance requirements. Of course, varieties of plastics only can be used as thermal forming. Plastic varieties that are often used to thermoforming include: various types of polystyrene, polymethyl methacrylate, PVC and styrene-butadiene-acrylonitrile copolymer, high density polyethylene, polypropylene, polyamide, polycarbonate and poly ethylene terephthalate two ester.

苯乙—丁二烯—丙烯腈共聚物

It's a method of manufacturing as a raw material for sheet of casting, rolling or extrusion.

Compared with injection molding, thermoforming is often with high productivity (the slot die production quantity per minute is up to 1,500 pieces), less equipment investment and can manufacture larger products and has other advantages, and the disadvantage is that the raw material's (sheet) high cost and product after processing. Therefore, whenever the two products can be produced, the choice depends on the production cost. It should be pointed out that, at present the development of two kinds of molding methods is faster, especially in thermoforming. Development situation can be changed now on the two kinds of molding methods for conclusion.

This chapter first describes the basic method, equipment and mold, then analyses and then briefly introduces the influence of process factors; finally, introduces a variety of commonly used thermoplastic sheet.

6.2 The Basic Methods of Thermoforming

热成型的基本方法

In order to facilitate production, we should change operation according to different product types. Thermoforming method can have a lot of changes (now about dozens). But the final analysis is always composed of several basic methods. Determination of the basic method is not consistent, but is essentially the same, different only in the subdivision degree. Here it will be divided into six types.

6.2.1 Differential Pressure Forming

差压成型

First, clamp the sheet with a holding frame in a mold and the mold is heated by heater. When the sheet has been heated to a certain temperature, put it away from the heater and use the appropriate measures to make the sheet on both sides with different pressure (Fig. 6 – 1 and Fig. 6 – 2), thus, the sheet will bend downward and fit with the mold surface. Followed by the fully cooled compressed air from the bottom of the mold through the vent it will blow molded sheet, and be trimmed to turn into products.

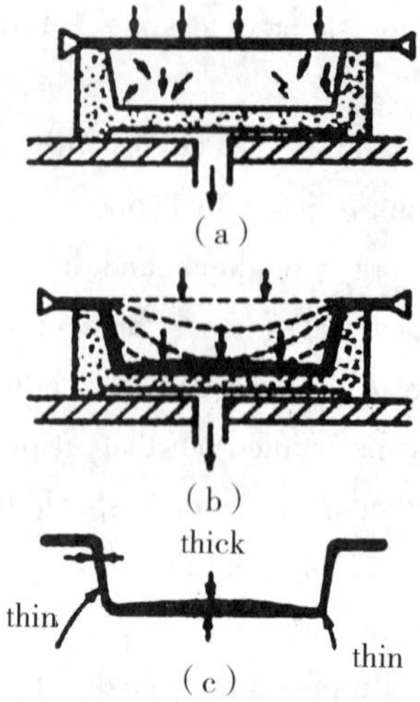

Fig. 6 - 1　Vacuum molding

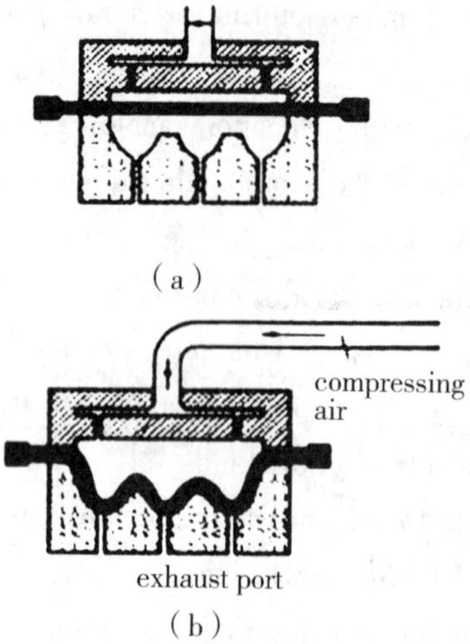

Fig. 6 - 2　Compression molding

真空成型
加压成型

There are three measures to make two sides of sheet generate pressure difference, evacuated from the bottom of mold, compressed air from the top of sheet, or both. The former is often referred to as vacuum forming, the latter two are known as compression molding. The maximum differential pressure the evacuation can cause is usually 0.07 - 0.09 MPa. If the pressure can not meet the requirements of forming, it should use compressed air pressure. Compressed air pressure is generally not more than 0.35 MPa. The pressure difference depends on the size of die and solid degree of equipment. Compared with the vacuum forming, compression forming not only can be used for thicker sheet production, but also can be formed under lower temperature, and has the advantage of short production period.

Differential pressure forming is one of the easiest thermoforming methods; the main characteristics of the products are that: the structure is clear; fine parts and die surface are in a joint; degree of finish is high; during the molding, when the sheet and the die surface are in fit time, its thickness is small. The thickness distribution is shown in Fig. 6 - 1(c).

抽空柜

自由成型

The mold for differential pressure forming is a single die, but we can use without mold. Without mold, sheet is clamped in the evacuated cabinet (vacuum forming) or flat plate with hole on (compression forming), which is shown in Fig. 6 - 3 and Fig. 6 - 4. When forming, only to a certain extent, you can stop time or pressure controlled by an intuitive or photoelectric cell. By this method (usually called free forming) production shape is hemispherical cover body; the surface is glossy without defects. If the plastic used is transparent, its optical properties can almost have no change, which is commonly used for the manufacture of aircraft parts, equipment cover and skylights . 仪器罩/天窗

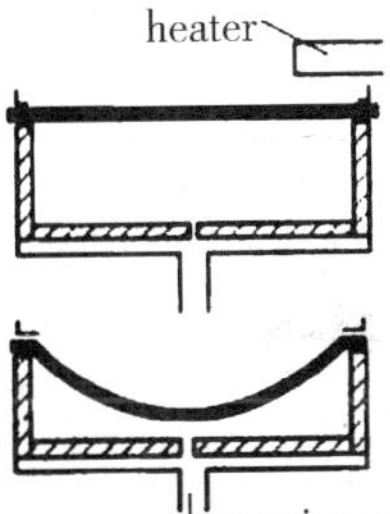

Fig. 6 – 3 Vacuum molding without model

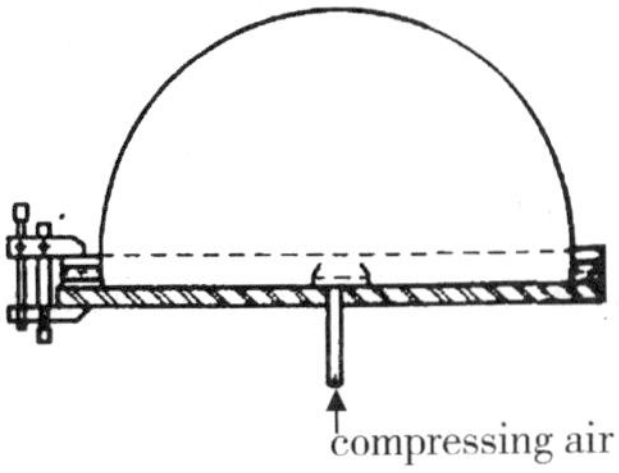

Fig. 6 – 4 Compressing molding without model

6.2.2 Cover Molding

覆盖成型

The molding is more favorable to produce the thickness and depth products. The mold is limited to a single male mold. When molding, it can use the thrust of hydraulic system to put male mold into the framework by clamping and heating plate(see Fig. 6 – 5) . It can also use the appropriate mechanical force mobile framework to put the sheet buckle on the mold. When sheet, framework and the mold have been completely fasten, extraction from the bottom of the mold makes sheet and die surface fully fit. After being cooled, products can be obtained after sintering and repairing. Products depend on the quality of the mold temperature, coverage rate, extraction rate, etc.

Cover molding products have the following main features. ① They are same as pressure molding, and die-face fitting of high surface quality is more explicit, detailed in structure. ② The thickest part of the wall is at the top of the mold, and the thinnest part is in the border area of the mold side and bottom, as shown in Fig. 6 – 5(d). ③The side face of products often arise drafting and the stripes of cooling. Because the time of different parts in the sheet sticking to the mold surface is different. And the part of the mold surface contact was first cooled, and in the process of subsequent buckle cover, its drafting behavior is weaker than the part that is still not cooling. This stripe is usually the closest to the side of the top surface.

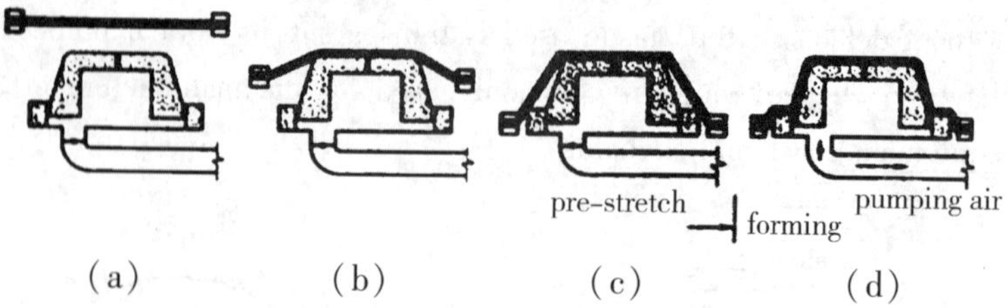

Fig. 6 – 5 Cover molding

柱塞助压成型

6.2.3 Plunger- and Pressure-Assist Molding

This shape is often divided into the pressure-assist vacuum forming and pressure-assist pressure forming, respectively as shown in Fig. 6 – 6 and 6 – 7. When molding, like other differential pressure forming, firstly, use clamping box to make the sheet tightly clamp on the female die, and heat the sheet with heater to achieve enough temperature. Then, in the case of closing die hand door, press the plunger into mold. When the plunger extends to the mold, due to a back pressure at the bottom of the sheet covers the plunger without contact with the die surface, the degree of piston pressure is limited to make the sheet stick to the bottom of the mold completely. When plunger stops falling, the plunger helps pressure vacuum forming evacuate from mold bottom to make the sheet and die surface completely joint. If pneumatic plunger helps pressure molding, plunger template is closely combined to die, and flux at the edge of the plunger into the compressed air causes the sheet and die surface to be fully fit. When the sheet has been forming, both methods lift the plunger into the original position. Sheet molding, cooling and sintering and after finishing, namely are becoming products. To a great extent, product depends on the quality of the plunger, the temperature of the sheet and the speed of the plunger downing. If plunger downing speed allows, the sooner, the better.

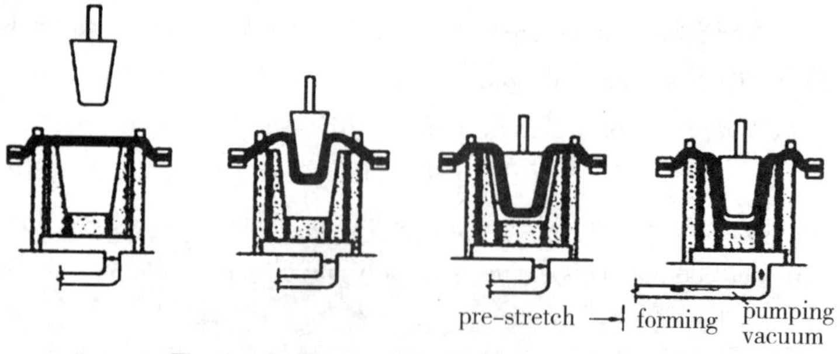

Fig. 6 – 6 Ram pressure-assist vacuum forming

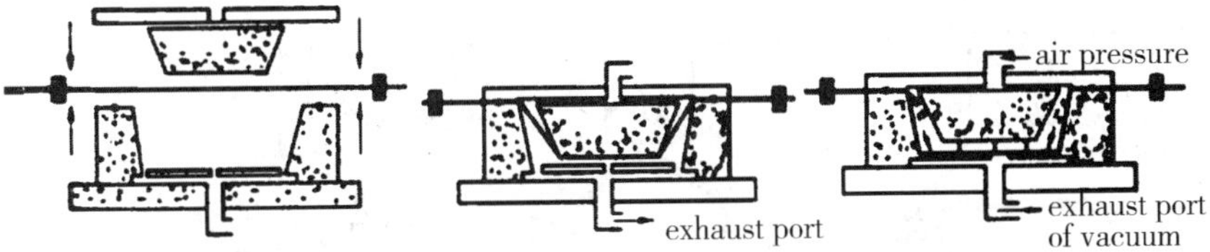

Fig. 6 - 7 Ram pressure-assist pressure forming

In the pressure-assist vacuum forming, when the piston is pressed into the mold, the sheet can slide in the die, such as enlarging the frame clamping sheet area in accordance with the die section, then we can put more sheets into the mold when plunger falls; the sheet thickness of the product is different and no amplification is smaller. Enlarging the extent to the specific shape of cavity depth and products prevails. By experience, to the general frame the distance clamping to die is 25% less than the depth of the mold slot. Cavity is shallow, causing the percentage to be smaller. Without amplification frame clamping sheet, that area and amplification of the sheet and the sliding clamp can also achieve the effect mentioned above, but the thickness can control measures such as amplification in the tight frame.

In order to make the thickness more uniform, before the piston downs, first with 0.01 - 0.02 MPa (or more) of the compressed air from the bottom of the mold, make the heating sheet bulging bubble appropriate; then in the control pressure cavity the piston downs; when the plunger will heat sheet drawing close to the bottom mold, stop falling, mold bottom pumping; when the sheet and the mold surface fully fit, finish shaping. This process is shown in Fig. 6 - 8; it is usually referred to as the inflatable plunger help pressure vacuum forming. If you don't use bottom pumping and the compressed air from the plunger end into the method of forming sheet, that is called flatulence plunger help pressure molding. By adjusting the temperature of the sheet, bubble level, the plunger velocity and temperature, the compressed air is applied and exclusion and the size of the vacuum degree can accurately control the thickness of products section, and the sheet can be uniformly extended. By this method the product thickness can be as small as the original film with thickness of 25%, and has good repeatability, usually without physical property damage to the original film, but the stress is one-way, so the products parallel to the stretching direction of rupture are sensitive.

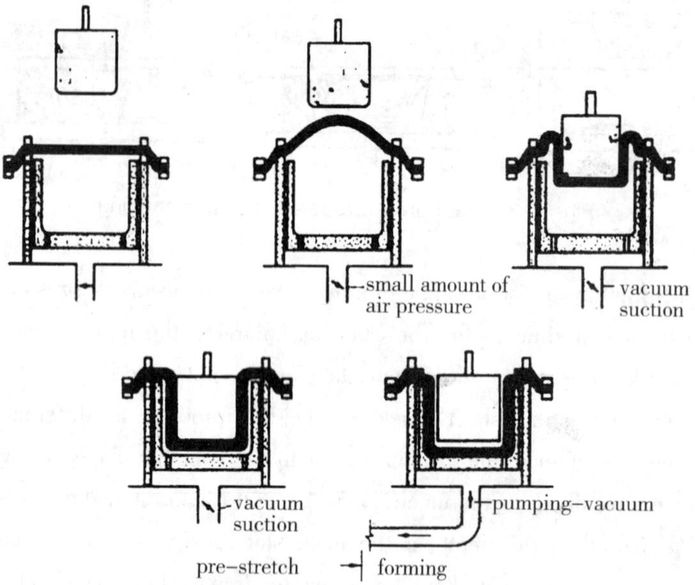

Fig. 6 – 8 Inflatable ram pressure-assist vacuum forming

Plunger helps press the volume to be 70% – 90% of the mold cavity volume, and usually the surface should be smooth, and generally with no fine parts in the structure.

In addition that thickness is uniform; the piston pressure-assist molding products' main features are very similar to other differential pressure forming.

回吸成型

6.2.4 Suction Molding

The commonly used are vacuum suction molding, inflatable gas vacuum suction forming and push gas vacuum suction molding.

Vacuum suction molding is shown in Fig. 6 – 9. With this method of molding, the first several steps, such as sheet clamping, heating, vacuum suction, and vacuum forming are the same as vacuum modeling. When the heating sheet has been sucked into the mold and reached a predetermined depth (photocell control), insert the upper of the mold into the sheet that has been bent, until the die edge is completely sealed in evacuated area. Then, open the valve to the area at the bottom of the mold and let it evacuate from the top of the mold. In this way, the sheet will fit the mold surface because of back-flow, and will get the products after cooling, sintering and dressing.

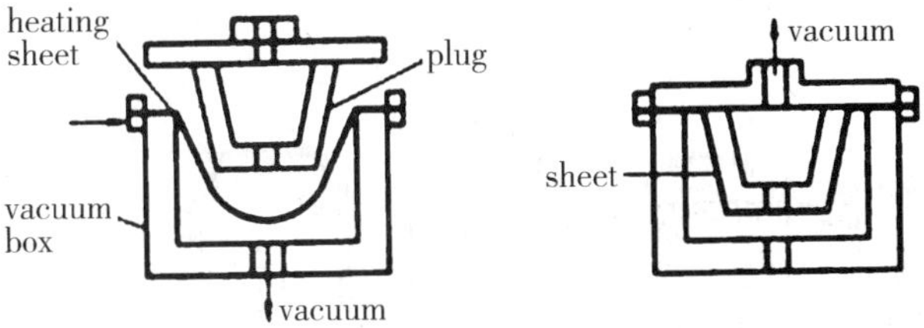

Fig. 6 – 9 Vacuum suction molding

Inflatable gas vacuum suction forming is shown in Fig. 6 – 10. The technology of sheet bending method is not by evacuating, but by compressed air. Compressed air is introduced from the bottom, which makes the heating sheet show the shape of bubble, then compresses the shape of bubble into the tank with column mold when it achieves the required height. During the process of the column mold extending into the pressure tank, the pressure inside should be maintained properly, and make use of the lower pressure of sheet at the bottom to wrap the column mold. When the mold extends to the proper part of the box and the edge of the mold is sealed in the evacuation zone, open the valve at the top of the mold to evacuate. Therefore, the sheet is back to the suction and can fit the mold surface, and the molding is completed. It will turn into products after cooling, sintering and finishing.

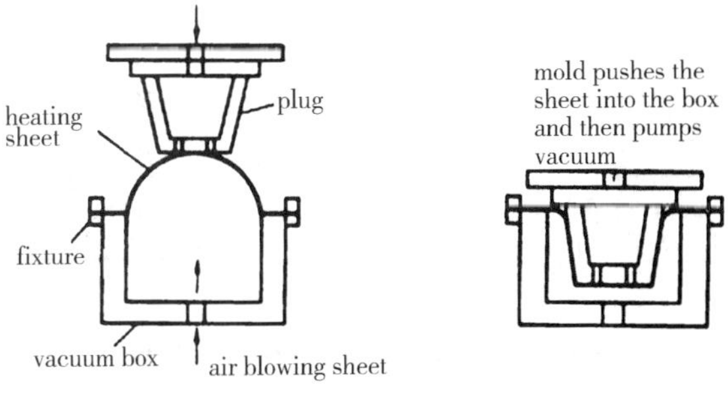

Fig. 6 – 10 Inflatable gas vacuum suction forming

Push gas vacuum suction molding is shown in Fig. 6 – 11. Molding sheet into bubble method is not using time and pressure on the edge and time zone for gas tight seal mold. When the mold proper positions have risen to the top, stop rising. Make the sheet at the bottom of the joint and mold, after cooling, sintering and dressing, which makes products. Back to the suction molding in products with more uniform wall thickness and structure is the main characteristic and also can be more complex.

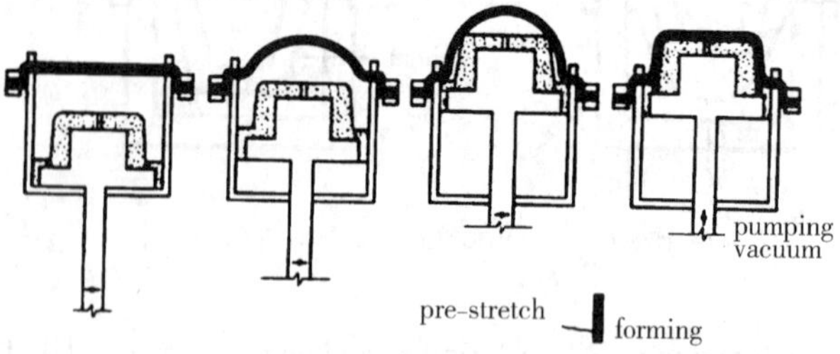

Fig. 6 - 11　Push gas vacuum suction molding

对模成型

6.2.5　Matched-mold Forming

Unlike the various methods above, this method, as shown in Fig. 6 - 12, adopts two single buckling molds, which are paired to mold. When molding, sheet frame clamping between the two molds and heated by a heater can be moved. When the sheet is heated enough, remove the heater, and put two molds together. The rise and fall of the upper and the lower mold are controlled by the hydraulic when fold, and during the fold process, the air between sheet and mold is discharged from the holes arranged on the mold. It will turn into products after cooling, sintering and fixing.

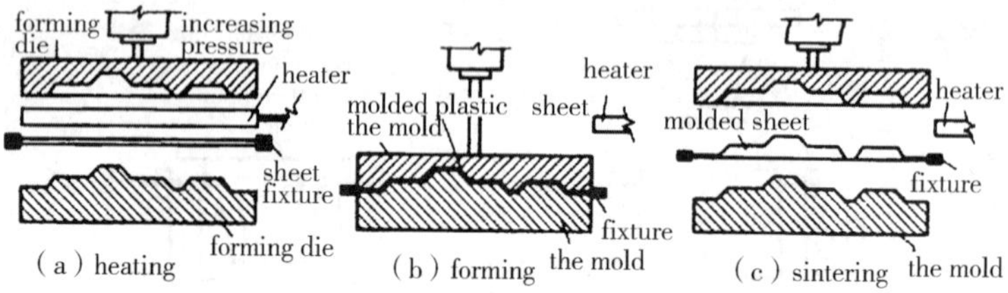

Fig. 6 - 12　Matched-mold Forming

In order to ensure the quality of the products, in the mold design we should consider the following two points: ①Don't make the thickness of product change with the pressure fluctuations between the upper and the lower mold. ② The upper and the lower mold must press the sheet plain. As the thickness change, it can be solved by setting up a fixed buckle point between the upper and the lower mold. The surface of the mold usually adopts the soft materials, such as foam rubber. The surface of the soft material should match the molds to some extent but is not so strict in the appearance.

Characteristics of molding products are that: ① Copy and dimensional accuracy are better; ② the structure can be more complicated, and even can be made into a surface engraving or carving; ③the thickness distribution, to a large extent, depends on the product.

6.2.6 Two-piece Thermoforming

双片热成型

This method is used to produce hollow products. The production process is shown in Fig. 6 - 13. When molding, put the two plastic sheets that have been heated to the required temperature in the frame of half mold and clap it. Then the needle inserted into the air will be compressed to the hollow area between the two sheets. At the same time, open the vacuum valve that is arranged when the two clamping is evacuated. Therefore, sheet is suitable for cavity in half mold. After cooling, sintering and dressing, it will turn into hollow products .

中空制品

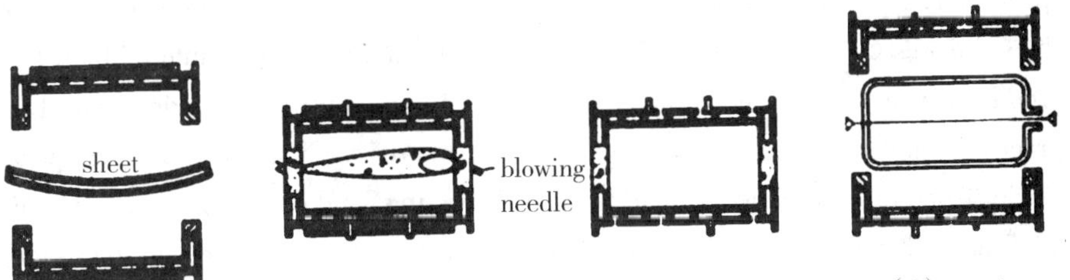

(a) two plastic sheets clamped on a cold frame (b) the introduction of compressed air from blowing needle (c) pumping vacuum (d) sintering

Fig. 6 - 13 Two-piece thermoforming

The characteristics of hollow products produced by this method are rapid prototyping, uniform thickness, color and different thickness of the products.

6.3 Thermoforming Equipment

热成型的设备

It is known from the above basic methods that each method includes five process: ① sheet clamping; ②sheet heating; ③forming; ④cooling; ⑤sintering. The forming is more complex; there will be great difference with different methods. For this reason, in the design of the device, except for few cases, all five steps can be completed. But according to the needs of production, only one or a few types can be formed.

There are manual, semi-automatic and fully automatic molding equipments. All manual equipment operations, such as clamping, heating, evacuating, cooling and sintering are completed by artificial adjustment. The operations of semi-automatic equipment, besides the clamping and stripping that must be completed manually, are set in advance according to the conditions and procedures by the device automatically. All the operations of automatic equipments are completed by the equipment automatically.

According to the feeding way, molding equipment is divided into partial feeding and continuous feeding. Partial feeding is used in the production of large parts, and the raw material used is generally the thick sheet that is not easy to roll. But partial feeding type of equipment can make the small parts with thin sheet. Partial feeding equipment commonly used in industry is of three rotary press. This kind of equipment is divided into three sections according to the loading and unloading, heating and forming process. Heater and forming mold are located within a fixed section; sheet is made up of three frames clamping according to the space of 120° angle that can rotate, and take turns to rotate in three sections, as shown in Fig. 6 – 14. Due to operation needs, the rotary motion is intermittent. For productivity or plastic need more heat, heating can also be divided into two processes, and become four rotary machines, but is not much used. There is a single period of molding machine in industry. Using this kind of equipment, in addition to discrete heating part, all other processes complete step by step in the same segment. The clamping frame's size of various molding machines usually can be changed within a certain range; the maximum size is determined by the equipment. Molding machine is used in the industry; its maximum size is 1.2 m by 1.8 m.

轮转机

Continuous feeding equipment is used for the production with thin wall of small parts, such as cups, plates, etc., and belongs to mass production. Such devices also have many sections; each section only completes a process, as shown in Fig. 6 – 15. Although the feeding is continuous, its operation is still intermittent, intermittent period ranging from several seconds to 10 seconds. To save energy and make material feeding more convenient, the equipment also has the sheet extruder feeding directly, as shown in Fig. 6 – 16. To shorten the process and make operation convenient, there is more material to be incorporated into the upper continuous feeding production line. During production, the packed material will be put into molding container, covered with plastic sheets; the hot sealing and packaging molding will be finished together.

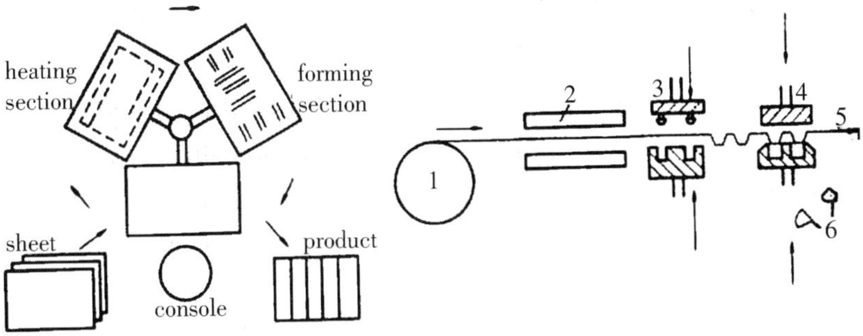

Fig. 6 – 14 Continuous feeding device processes

(1—film reel;2—heater;3—mold;4—blanking die;
5—piece mold material recycling;6—product)

Fig. 6 – 15 Extruder feeding continuous molding equipment

Forming machine adopting sheet extruder feeding is usually not directly attached to the heating section. Because the feeding of extruder is continuous, but the molding is intermittent, it should be taken into account in the molding machine structure, in order to overcome this contradiction. The majority is to design the mobile equipment in the forming section, so that it can be moved forward with the extrusion sheet in a certain position of every process; but when it completes a work cycle, forming section can immediately go back to the original position for the next cycle of operation. It's more favorable for mass and long-term production by using extruder molding, but there are still two main drawbacks: the extrusion rate is generally faster than the thermoforming rate, so that each production cycle is often longer than other molding machines; control factors of production are various, and working of each section has the synchronization requirements, which will make the production management and maintenance causing a bigger trouble.

The design of several main parts of all kinds of equipment will be different based on the different requirements. The general situation and main requirements are as follows.

加热系统

6.3.1 Heating System

The heating method of the sheet is not strictly limited; it can be the conduction of hot plate and also be a infrared irradiation. The supply methods of hot plate heat are electricity, hot water, steam, etc. Thicker sheet usually uses another oven preheated, so as to reduce the machine's load and increase productivity.

Almost all modern molding machine heaters adopt infrared irradiation. Infrared radiation efficiency depends on: ① the temperature of the heater; ② the density of irradiation; ③ the distance between the sheet and the heater; ④ the radiation absorbing performance of the sheet. Therefore, of all the equipments, in addition to the ④ , the rest all have certain regulation and are attached with the accurate effective adjustment and control equipment. Adopting infrared irradiation has an advantage of a shorter heating time, higher heating speed, energy saving, etc. Using the electric wire, it should be throughout the porcelain casing, and be made into the covered heater, which is safer. The temperature of the heater generally is 370 ℃ – 650 ℃, Power density is about 3.5 – 6.5 W/cm^2. Heater temperature is very high, but the sheet is not directly connected with the heater when using. It is heating indirectly in mode, and the distance between the heater and the sheet are adjustable, in order to control mold temperature. The distance ranging between the sheet and heater is 8 – 30 cm.

In order to improve the heating rate and the production efficiency, when the sheet thickness is greater than 3 mm, it often needs to use two-side heating methods, which is equipped in each of the upper and lower clamping sheet with a heater. When both sides are heated, the temperature of the lower heater should be lower than the upper heater, because the hot air increases, and thus also can prevent the sheet from sagging excessively in heating process. If the rotary press or continuous feeding device works, it can also use multiple sections of heating to achieve the same purpose, but each section of the heating conditions is individually controlled and requires strict coordination between the sections. In single segment molding machine heater and the rotary press or continuous feeding of different equipment, it is the intermittent work and can be temporarily removed after heating. Heater power is not truncated after removed, but makes good use of the reflector or heat insulation bed to keep it in high temperature.

反射器

Heater heating surface of the clamping area is usually slightly larger than clamping frame. The former edge over the latter line length is 10 – 50 mm.

In the design of heating system, the plastic heating power density is best given. Several kinds of commonly used plastic heating power density are shown in Table 6 – 1.

Table 6 – 1 Commonly used plastic heating power density

Plastic	Power density(W/cm^2)
polyvinyl chloride(PVC)、modified polystyrene	1.5 – 3
polyethylene、polypropylene	3.5 – 5
PSU/PC	≥5

The first cooling molding of products should be done as soon as possible after completion, so as to improve productivity. The cooling methods are internal cooling and external cooling. Internal cooling makes the primary product cool through the mold cooling. Internal cooling is only limited to metal mold. External cooling is air cooling method (the fan) or air-water spray method . The spray cooling is rarely used, because it is easy to make products produce cold scar, and bring the problem of dehydration. 空气—水雾法

Mold temperature is usually maintained at 45℃ – 75℃ during molding. If the mold is metal, the warm water circulation can preset on the die channel. As to the non-metal mold, it can only use the method of hot and cold to maintain the temperature due to the poor heat transfer. Use the infrared radiation to heat, and cool with air cooling. In the auxiliary plunger molding, in order to prevent the cooling sheet for piston causing molding to stop, plunger must remain the same or similar temperature with the sheet. Based on the requirements of the application, piston should be made by metal, in order to make the method of internal heating maintain proper temperature. 非金属模具

6.3.2 Clamping System 夹持系统

Clamping frame is composed of two racks. The upper rack is controlled by compressed air, often pressing the sheet in the bottom rack with balance. Pressure can be adjusted within a certain range. The total pressure rises while the size of equipment changes, which is from 0.5 t to 5 t. Clamping pressure generally will not become uneven as the sheet thickness is different, so the frame is often accompanied with automatic pressure compensation device. Equilibrium of the clamping is the necessary condition to guarantee the uniform distribution. Clamping sheet should be with reliable tightness, otherwise, it will cause leakage, 夹持框架

sheet sliding(the permitted sliding is another matter, and the device used is not the same), torsional and bending of sheet, etc. when molding.

水平方向 To meet the needs of production, holding frames are mostly made to be moved in vertical or(and) in horizontal direction .

真空系统

6.3.3 Vacuum System

抽空系统

叶轮式

The equipment that needs thermoforming generally is accompanied by itself. Pump equipment, is only used in the factory which has more equipments—using centralizing pump system. Self-supporting equipments using vacuum pumps are mostly the impeller type. The vacuum degree it can achieve is usually 5,096 Pa, and under special circumstances is 6,566 Pa. The rotating power of the vacuum pump is determined by the size of the molding equipment. The large equipment is usually 2 – 4 KW. The size of centralizing pump system is determined by the factory production and development of the specific requirements, usually is designed according to predetermined planning. Adopted vacuum pump types have impeller type, steam injection type and so on. It has to set up vacuum zone to stabilize the pressure when pumping the air, either self-sufficient equipment or centralized system.

蒸汽喷射式

压缩空气系统

6.3.4 Compressed Air System

额定容量

储压器

Compressed air systems are mostly used in molding. Compressed air system also has self-sufficiency and concentration in two forms under specific circumstances. The air compressor attached in the medium-sized thermoforming equipment mostly has single stage or two stages. Its rated capacity is 0. 15 – 0. 3 m^3/min; pressure range is 0. 6 – 0. 7 Mpa. Any compressed air system should be attached to storage pressure gauges to stabilize the pressure.

模具

6.4 Mold

石膏

酚醛树脂

As the pressure of the thermoforming is not very high, the mold rigidity requirements are low. Therefore, the material of molding is not limited to steel. Common molding materials are wood(hard wood and compression wood), plaster, plastic(phenolic resin , epoxy resin, polyester resin and amino resin), aluminum and steel. The selection of mold materials is mainly based on the production quantity and quality. When producing trial products, it generally uses the hardwood or plaster mold. When it has little yield using the plastic, high yield or high quality (the surface degree of finish is high or the structure is fine) can use aluminum or steel. In addition, it can also use the combination mode of wood, plastic and metal.

As the foregoing, when adopting the single mode shape molding, the part of surface with high quality of products is the side contacting the mold surface, and it is more distinct and detailed in structure. Therefore, the rule of selection of the female mold and male mold is in that. When using multislot mold, it had better use female mold, because the interval between the cavity can be more compact, and at the same time it can also prevent the sheet in the molding process and mold surface wrinkling when in contact. In addition, the releasing of female mold is easier, but the products section at the bottom is thinner.

There is a close relationship between the mold surface degree of smooth and the product surface. Highly polished mold will make products' surface glossy, and stuffy light mold will make products' surface radiant. But others are dull. However, all kinds of plastic have their own thermal strength, tensile strength and the bonding characteristics of mold surface, then the requirements are not the same. For example, polyolefins with smooth mold surface forming is difficult, so the surface of mold for the plastic thermoforming should be properly rough. 高度抛光模具

Mold spiral groove depth and width ratio are often referred as the draft ratio, and it is an index to distinguish the quality of thermal forming methods. Generally speaking, drafting ratio can be larger when using a single male mold molding, because it can use sheets to tow or pre-stretch by force, but the draft ratio is not larger than 1. The drawing ratio of using female mold is usually not more than 0.5. The purpose of plunger pressure assisting molding is to drag and drop or pre-stretch sheet, so the draft ratio of this molding can be higher than 1. 牵伸比 指标

In order to avoid the formation in stress concentration, and improve the impact strength, the edge and corner angle on the surface of the mold should not use the sharp one but fillet. The radius of the fillet had better be equal to or bigger than the sheet thickness, but not less than 1.5 mm. Mold wall should be set for sintering inclination. The female mold angle is 0.5° – 3°, and the male mold is 2° – 7°. The pore diameter is set in the mold to import gas when forming in the mold, and its size is slightly different from the type and thickness of sheet. Suppression of soft PVC and polyethylene sheet is about 0.25 – 0.6 mm; the other sheet is 0.6 to 1.0 mm; as for the hard and thick sheet it can be up to 1.5 mm. When aperture is too large, it will often make the product surface have instrument. 冲击强度

Considering reducing gas by stomatal resistance and easily molding, change the end hole near the bottom of the mold into a large diameter, and its size is 5 – 6 mm, or even adopt a long and narrow slit. The place of vent hole is mostly in depth of the corner, the center of larger flat and partial concave part. When in the fine parts, the distance of vacuum hole arranged in can be close to 6.4 mm; the plane mold's vacuum hole distance can be 25 – 76 mm.

The producing method of vent hole is determined by the material of mold. The mold that is casted with gypsum, plastic and aluminum, has to set the vent hole by placing tiny copper wire usually in the casting process. And it has a hole after finishing casting process. Wood, metal, electrolytic copper and other mold need drill to punch. For smaller diameter holes, firstly it is with coarse drill bit to drill to the surface from mold with 3 – 5 mm, then with small diameter drill to punch again.

The shrinkage rate of thermoforming products is about 0.001 – 0.04. To acquire the accurate size of product, shrinkage rate will be given appropriate consideration when designing the mold. Several kinds of plastic thermoforming shrinkage rate are shown in table 6 – 2.

Table 6 – 2 Several thermoforming shrinkage rate of plastic

Plastic	Shrinkage rate of product	
Polyvinyl chloride (PVC)	Male mold	0.001 – 0.005
	Female mold	0.005 – 0.009
ABS	Male mold	0.001
	Female mold	0.004 – 0.008
Polycarbonate	Male mold	0.005
	Female mold	0.008
Polyolefin	Male mold	0.01 – 0.03
	Female mold	0.03 – 0.04
Modified polystyrene	Male mold	0.005 – 0.008
	Female mold	0.008,9 – 0.01

6.5 Technological Factors Analysis

工艺因素分析

Although thermoforming operation is simple, the process involves a lot of factors. To get satisfied products, it is necessary to understand these factors. Here is only thermoforming of sheet heating, shaping and cooling mold releasing products involving in the main technological factors being briefly analyzed. Finally illustrate how to calculate from the specifications of the products design and calculate the forming of the specifications of the sheet thickness (batch feeding type).

6.5.1 Heating

The required time of molding temperature is 50% – 80% of the molding cycle. Therefore, how to shorten the heating time has its important significance. Heating (or cooling) time increases when the thickness of sheet and specific heat capacity increase, and decreases with the thermal conductivity and heat transfer coefficient, but it is not a simple linear relationship. For example, under the same conditions, if heat the sheet with different thickness, the specific relationships are shown in Table 6 – 3.

传热系数

Table 6 – 3 The relationship between heating time and the thickness of the sheet (polyethylene)

The thickness of the sheet (mm)	0.5	1.5	2.5
The time required for heated to 121 ℃ (s)	18	36	48
Heating time of unit thickness (s · mm)	36	24	19.2

Based on the above reasons, the sheet adopted should be uniform thickness, or will make the emergence of the phenomenon of uneven temperature, so that the product will generate internal stress. Thickness tolerance of plastic sheet is generally not more than 8%, otherwise it should extend the heating time, and let the heat transmit to the thick wall inside, and make sheet "heat through" and have uniform temperature to ensure the quality of the products. Because plastic has small thermal conductivity, when heating the sheet, if the larger power heating device has adopted or the heating device is too close to the sheet, the top of the sheet will reach the desired temperature, but the temperature of the bottom surface will still be very low. While the bottom surface temperature is reaching the requirement, the top surface temperature has been exceeded, or even has been burned. The low temperature of the bottom surface and the high temperature of the top are not suitable for molding. In this case, the best methods are double-side heating, pre-heating or high-frequency heating (if the sheet can use this method to heat) to shorten the heating time. The lowest temperature of molding should be based on the sheet that is not pale or appears obvious defects when drafting, and the maximum is that the sheet will not become degradation or unduly droops in the clamping frame. In order to obtain the fastest molding cycle, forming temperature is usually partial to the low limit value. For example, the lower limit of the molding temperature of styrene-butadiene-acrylonitrile copolymer (ABS) can be as low as 127 ℃, and the high limit is 180 ℃. Only when rapid vacuum forming drafts lowly articles, can molding temperature be 140 ℃. And products deep drawing temperature is about 150 ℃. Only when products are near to the higher limit, can it be about 170 ℃. Sheets appearing saggy frame are thermal expansion and the melt flows in the folder. Thermal expansion of the sheet is the phenomenon of being heated from room temperature to the temperature of all non-directional. In this case, each sheet's growth in the size is 1% – 2%. The size of melt flow depends on plastic's melt viscosity. Sagging phenomenon often can be overcome by using a sheet orientation. The directional sheet has heat shrinkage behavior. But too many oriented sheets often make the molding difficult because of too much stress.

The sheet drawing of each part changes according to the different mold pattern when molding. In order to prevent the parts making different thickness, drafting is relatively strong when covered with specially shaped paper cut ceiling, and makes these parts reflected by less infrared irradiation, so that the temperature here is a little lower. In this way, it can make the thickness of the products a little better uniform. From the experiment, this kind of products, because of the relation of internal stress, in terms of dimensional stability and mechanical properties, will be affected. The general performance is the covered part of the smaller dimensional stability and high impact strength. Improving overall mold temperature can often reduce the internal stress of products and achieve better dimensional stability. In addition, the mold and auxiliary piston should use the appropriate temperature according to the different plastics, and make the products have good quality. All kinds of plastic sheet molding conditions and thermal expansion coefficient are shown in Table 6 – 4.

因次稳定性

Table 6 - 4 Molding conditions and thermal expansion coefficient of various kinds of plastic sheet

Types of plastic	Forming conditions				Thermal expansion coefficient [$\times 10^{-6}$ cm /(cm · ℃)]
	Mold temperature(℃)		Mold temperature (℃)	Auxiliary piston temperature (℃)	
	Maximum value	Minimum value			
Rigid polyvinyl chloride	135 - 180	93 - 127	41 - 46	60 - 149	6.6 - 8
Polyethylene(low density)	121 - 191	107	49 - 77	149	15 - 30
Polyethylene(high density)	135 - 191	121	64 - 93	149	15 - 30
Polypropylene	149 - 202				11
Polystyrene(biaxial orientation)	182 - 193		49 - 60	116 - 121	6 - 8
Styrene-butadiene-acrylonitrile copolymer	149 - 177	140 - 160	72 - 85		4.8 - 11.2
Polymethyl methacrylate (injection)	143 - 177				
Polymethyl methacrylate (extrusion)	110 - 160				7.5 - 9.0
Cellulose acetate	227 - 246	216	77 - 93	274 - 316	7
Polyamide 6	216 - 221	210			10
Polyamide 66	221 - 249				10
Polyethylene terephthalate oxalicester (orientational)	177 - 254				
Polyethylene terephthalate oxalic ester(undirected)	177 - 204				

6.5.2 Forming

成型

When molding, the major cause of the uneven thickness of sheet is the different drawdown between different parts. Although the problem of uneven heating can be solved by heating with ceiling and the reasonable distribution of vent hole on the molds, these kinds of methods will decrease the dimensional stability of products. The uneven thickness caused by different drawdown is not the same in the various molding methods, of which the most serious method is differential pressure. Another factor affecting products thickness is drafting or sheet dragging speed, which is the rate of extraction, reassuring, or mold, framework and auxiliary pistonmovement rate. Generally speaking, the rate should be as fast as possible so that it is good for molding itself and shortening the cycle, and sometimes even changing the hole into a long and narrow seam. For example, when a refrigerator bush has been bush molded, if we use normal vacuum hole, the time should be 2 – 5 s; after using a long slit, the extraction time can be reduced to 0. 5 s. Of course, too fast rate will make the partial concave or convex parts' thickness too thin due to insufficient flow rate of the products. However, too slow rate will appear cracks due to the sheet of the first cooling. Drawing rate depends on the sheet temperature, therefore, the thin sheet drawing is generally faster than the thick type, because the former temperature will drop faster when molding. Deep drawing pro ducts depends on suction or blowing into even stretch vesicular shape before molding, in order to ensure the uniformity of the products section. The thickness of the bubble shape should be close to the required thickness of the bottom of the products. Because the molding of mold or plunger contacts with plastics, the drawing is very low. The bubble of specified requirements shall be maintained for a few seconds and let it to adjust the shape itself, homogenize the section and make the hotter (thicker) part of the sheet to continue, and receive the slight contraction of cooler (thinner) portion.

电冰箱

If draft is uniform in all directions when molding, the performance of the products and each upward are no difference. But the drawing is hard to meet in the real situation. If drawing of molding lays particular stresses on one direction, the products will appear anisotropy because of the molecular direction. Anisotropy usually performs in terms of mechanical properties. For example, tensile strength, elongation, crack resistance, etc. will be increased, while in other directions will be relatively weakened. The stability of molecular targeted product is weaker, especially in the case of heating. Production in practice shows that under the correct mold temperature, if the difference between unidirectional draft numerical and two-way drafting remains within a certain range (specific values vary with plastic varieties), the degree of anisotropy of the products is not very large. To clarify, improving the forming temperature and reducing drawing rate have the effects on reducing the importance of molecular orientation.

抗裂能力

6.5.3 Cooling and Sintering

冷却脱模

Because of poor thermal conductivity of plastics, with the increase of the thickness of calendering mold, cooling time will increase, and has to shorten the molding cycle. As mentioned before, the cooling portion has divided into internal cooling and external cooling, and they can be used not only independently, but also in combination. It should be based on product's needs. It usually adopts external cooling, because it is easier. No matter what kind of way of cooling is adopted, it is important to make the molding products to release below the deformation temperature. For example, PVC is 40 ℃ - 50 ℃; cellulose acetate is 50 ℃ - 60 ℃; PMMA is 60 ℃ - 70 ℃. If cooling is insufficient, products will change the shape after sintering, but excessive cooling in some cases will make it difficult to release, for the shrinkage side of the product ties to the mold.

醋酸纤维素

In addition to the excessive heating of plastic sheet that causes decomposition or makes the mold surface too rough, few sheets adhere to the mold. If this kind of phenomenon appears occasionally, the release agent can be applied on the surface of the mold to remove the defects. But the amount of using should not be too much, because it will affect the smoothness and transparency of the product. Common varieties of release agent are zinc stearate, MoS_2 and organic silicone toluene solution.

二硫化钼

6.5.4 Sheet Thickness

片材厚度

There is a lot of molding processes and product types, so it is difficult to be careful about the calendering mold's specific changing in the thickness of various parts of products. The correct choice of the molding sheet thickness and area depends on the product actual section's minimum thickness and shape. Here are the concepts from the area ratio calculation of sheet thickness. Area ratio is associated with the drawing depth, and it can be shown with available product area and original area ratio of the sheet. For example, if the sheet area is 300 cm^2, and the total area of the product is 600 cm^2, the area ratio is 2:1. Products' thickness is half of the sheet thickness, therefore, knowing the area ratio can estimate the sheet thickness roughly. We present two cases of calculation in the following.

Example 1: Using cover molding method, product size is 610 mm × 500 mm × 50 mm (length × width × depth); thickness is 2.3 mm; an area of 416,000 mm^2 was covered by the materials used for clamping and mold margin width is a total of 20

覆盖成型方法

mm; try to calculate the size and thickness of the sheet.

Solution: (1) The size of the forming sheet.

Length is $610 + 20 \times 2 = 650\ mm^2$.

Width is $500 + 20 \times 2 = 540\ mm^2$.

(2) The thickness of the calendering mold.

有效面积

Forming sheet of the effective area is $610 \times 500 = 305,000\ mm^2$.

Ratio of the area is $416,000/305,000 \approx 1.36$.

So the sheet thickness is $2.3 \times 1.36 = 3.128$ mm.

Example 2: The size of the product that gets from the plunger assisted compressed air expansion molding is 1,270 mm × 610 mm × 500 mm (length × width × depth); its thickness is 2 mm; an area of 2,654,700 mm^2 was covered with the width of the sheet clamping frame between the mold and a total of 12%; clamp clamping sheet is 20 mm; the width of a piece of material in contact with the plunger uses 100% as the bottom of the product, while the rest of the material is using only 50%; what about the molding sheet size and thickness?

Solution: (1) The size of the sheet.

Length is $1,270 + 1,270 + 20 \times 2 \times 12\% = 2,544.8$ mm.

Width is $610 + 610 \times 12\% + 20 \times 2 = 723.2$ mm.

(2) The sheet thickness.

The bottom area of product is $1,270 \times 610 = 774,700\ mm^2$.

The length of the sheet removing clamp clamping is $1,462.3 \times 2 = 2,924.6$ mm.

The width of the sheet removing clamp clamping is $723.2 - 20 \times 2 = 683.2$ mm.

The rest of the area is $(1,422.4 \times 683.2 - 774,700) \times 50\% = 98,541.84\ mm^2$.

The effective area of sheet is:

$774,700 + 98,541.84 = 873,241.84\ mm^2$.